Automation and Robotics for Farmers

THE AUTHOR

Dr. B. Sudhakar working as an Assistant Professor (Agricultural Extension), Department of Agricultural Extension, Faculty of Agriculture, Annamalai University, Chidambaram, Tamil Nadu. He did M.Sc (Agri.) in Agricultural Extension from Annamalai University and completed his Ph.D in Agricultural Extension from TNAU, Coimbatore. He has qualified NET (National Eligibility Test) exam conducted by ICAR twice and attended ARS (Agricultural Research Scientist) interview. He has more than 19 yeas of experience in Teaching, Research and Extension activities. He has published 4 text books and 30 Research papers in National and International Journals of repute. He has attended various National and International seminars and conferences and presented research and policy papers.

Automation and Robotics for Farmers

— *Authors* —
B. Sudhakar

2023
Daya Publishing House®
A Division of
Astral International Pvt. Ltd.
New Delhi – 110 002

ISBN: 9789359190921 (Int. Edition)

Published by : **Daya Publishing House®**
A Division of
Astral International Pvt. Ltd.
– ISO 9001:2015 Certified Company –
4736/23, Ansari Road, Darya Ganj,
New Delhi-110 002
Ph. 011-43549197, 23278134
E-mail: info@astralint.com
Website: www.astralint.com

Preface

This textbook have give detailed information on **Automation and Robotics for farmers** through which the contents like **Definition of Automation and Robotics –** Different authors quoted definitions and discussed. **Different Types of Automation and Robotics.**

Next about **Robotics Application in Agriculture and Farmers.** Highlights the **IoT and Wireless System Networks in Agriculture Operations, Private Companies Technologies in Automation and Robotics in Agriculture and Robots in Farming Operations.**

Next highlights the different **Robots in Farming Operations, Sensors and Automation and Success Stories of Startup in India on Agriculture Automation and Robotics.**

Next I briefed with **20 Robots Revolutionizing the Agricultural Landscape, Agricultural Robotics Marketng in the World, Enabling Technologies for Future Robotic Agriculture Systems and Human-Robot Interaction.**

Next I highlited some positive points sudh as **Benefits of Automation and Robotics in Agriculture and Advantages of Agricultural Robots.**

In the final section I discussed about some negative aspects such as **Agricultural Robots Disadvantages and Challenges of Farm Automation and Robotics in Agriculture.** At last I discussed about the **Future Scope for Automation, Robotics and Wireless System Networks in Agriculture.**

I trusted that this text book will be useful to the researchers, scientists, policy makers, students researcher, extension functionaries, both central and state government research stations and research centers of agricultural and allied subjects, universities and colleges, NGO's, private agencies and farmers (who is the target group for development and dissemination of technologies). All efforts have been made to make the book most up-to-date.

B. Sudhakar

Contents

Chapter 1

Introduction

The contribution of agriculture plays an important role in the economic growth of India. In India, more than 60 per cent of peoples depend upon agriculture. This 60 per cent of people produce food for the country. In India, recently the number of farmers cultivating crops are decreasing day by day. The agriculture results depend upon the various controlled and uncontrolled factors like sudden change in climatic condition; high and low rain fall and also the timing of the rain, storms and heat waves, and effects of cold *etc.* For these conditions, traditional farming techniques are not appropriate. To minimize these effects and increase the productivity of agriculture, automation and robotics techniques are required. By the help of automation and robotics, we can measure the environmental condition and water management because these are the most important factors in agriculture (Sirisha and Somasekhar, 2019 and Lakhwani *et al.,* 2019).

Agricultural production system has witnessed a paradigm shift in last few decades with advancements in precision agricultural technologies. Expanding world population and shrinking landholding size has italicized the researcher's to turn up towards application of IoT (Internet of Things), Artificial intelligence, machine learning and robotics for economically viable, environmentally safe and sustainable agriculture. Amelioration in the field electronics and computer science has provided opportunities to the experts to explore the possibilities to develop more rationale and adaptable robotic system based on external environment response. Various autonomous remotely controlled prototypes have been developed for orchards and horticultural crops such as oranges, strawberries and tomatoes. Robotic automated systems with camera and GPS navigation has also been tried in weeding, spraying and harvesting operations. In addition to that, robotic automated systems have also been explored in automated feeding and cleaning in indoor environment of animal production system. In India, a robot has been successfully employed in industries, service sector and security. Reviews suggest that very few attempts have been made to explore the possibility of robotics and mechatronics in agriculture. Research institutes like, TNAU (Tamil Nadu Agriculture University), IARI (Indian Agriculture Research Institute) New Delhi and CIAE (Central Institute of Agriculture Engineering) Bhopal has attempted few prototypes in seeding and planting, whose field viability is yet to be tested. Therefore, they provide an enormous appertain to precision agriculture scientists to explore the possibilities to enhance the productivity and safeguarding the environmental concerns with use of robotic technologies. This approach may not be economically justifiable to replace expensive repetitive labor. Robotics in crop production cycle could be reduced to three stages: Seeding, Plant care and Harvesting.

The future of agriculture production system aims to produce high quality and quantity agricultural produce for wide range of consumers. The main challenges to agricultural researchers around the world to be perceive sustainable agriculture by judicious use of ever limiting natural resources and prevention of environment. Farm mechanization is one of the determining factors for advancement of agriculture production system and to feed the burgeoning human population. This has stressed the inventors around the world for development of modern machineries and agricultural equipment's to perform and understand the complex agricultural processes and execute all the required operation with high precision, efficiency and minimal damage to the environment. To control and regulate these machineries during field operations information has to collect with the sensing element and to be stored in the data acquisition system for further processing. Therefore, the response of machinery to the external environment is one of the important criteria for efficient application of these advance machinery. The operation, care and maintenance of these machineries require a well skilled labor and computing expert. Various researchers around the globe has applied precision agriculture techniques to increase the yield of the crop and quality while reducing the input agriculture cost (Nof, 2009, Zhang *et al.*, 2013, Tremblay *et al.*, 2011 and Tremblay *et al.*, 2010).

In addition to the exiting precision agricultural practices, the studies indicate that the use of agricultural robots and autonomous tractors has reduced the drudgery of farm operation and input investment cost of farming (Gonzalez-de-Soto *et al.*, 2016 and Gonzalez-de-Soto *et al.*, 2015).

In advancement of technologies in 20th century has forced the developed countries to shift from traditional and conventional agricultural practices to advance agriculture (Ceres *et al.*, 1998). The automation, mechatronics and robotics has significantly increased the productivity in agriculture by increasing the efficiency, efficacy, reliability, precision and reduction in human intervention (Schueller, 2006). Despite of all these improvements in current agricultural practices, fragmented land holding, decreasing farm labour and climate change possess a demand for efficient farming solutions, mainly in developing countries. Therefore the demand of intelligent agriculture machinery is of utmost important to mitigate the above mentioned problems in agriculture. Robotics is an intelligent machine created by engineering marvel of computing, electronics and mechanical engineering which resembles the work pattern of human being.

Robotic technology in agriculture has for various operations like seeding, planting, spraying, weeding, harvesting and post-harvest operations (Xia *et al.,* 2015). Robotic technology not only removes the drudgery in farm operations, but also protects the farmers from work under harsh environmental conditions (Noguchi *et al.,* 2004 and Zhang *et al.,* 2013). Though, robotics have increased the attractiveness towards farming professions, non-repetitive task and complex variable working environment in agriculture makes it very difficult for application (Hiremath *et al.,* 2014 and Nof, 2009).

Furthermore, agricultural production deals with live produce (fruit, vegetables, grains and flowers) which is highly sensitive to environmental and physical conditions (Eizicovits and Berman, 2014), such as temperature, humidity, gas, pressure, abrasion and acceleration. Live produce requires gentle, accurate and often complicated handling operations to maintain sufficient quality to travel the distance and time separating their production site from consumers. This characteristics makes the replacement of human ability by machines or automation extremely difficult. Therefore, most fruit, vegetable and flower growing and similar production tasks, *e.g.*, trellising, harvesting, sorting and packaging, are still performed manually (Zion *et al.,* 2014). This makes manual labour a major cost component in field operations (Eben-Chaime *et al.,* 2011), reaching up to 40 per cent of the total cost.

Therefore, application of robotics in agriculture is a complex process, which requires expert system in dealing with variable and unstructured environment and does the intended work. In last few decades researchers from various universities and tried to develop robotic system for various agricultural operations such as seeding, planting, weeding, spraying, harvesting and material handling of agricultural robots. Most of these works has been carried in control environmental and laboratory in different countries. Hence there is need of extensive research for application of robots in real field conditions (Bikram Jyoti *et al.,* 2020).

Automation can be effectively used in food safety, to reduce inappropriate use of chemicals, to protect the environment, to maintain occupational health and in animal welfare. The record of treatment and steps carried out during the production is needed so GPS (Global Position System) is attached to machinery and origin of crop is also obtained. Automation is also used to find soil organic matter, moisture content in it and nutrient availability in the soil. Portable sensors can be used for measurement of several specific locations for agriculture. The robotic weed control system is used to identify weed. Recognition of diseases and pests is made by microbial sensing technology. Automatic record of the condition of crop is obtained by automated machines which can be viewed in different locations by using the software. By this, we can make environment-friendly agriculture (Baerdemaeker, 2013).

The study about the design of a task-based modular reconfigurable agriculture robot, the population in the future grows at a faster rate. There might be a shortage of food and agricultural land also decreases. People prefer to work in comfortable jobs rather doing agriculture. So the solution to this problem is using robotic manipulators. In today's technology, we have single task targeted robots. Creating a universal robot by integrating all the capability to a robot is expensive and complicated. To make different tasks a robot should be reconfigurable. For example, the harvesting task requires both accuracy and large workspace. 3 degrees of freedom *i.e.*, one prismatic and two revolute models can be used to harvest both peach and tangerine. By adding prismatic joints and allowing the order of the revolute and prismatic joints together we can increase the flexibility of the robot. Another approach, multitask problem can be optimized by using different robots for different tasks by using the same software (Levin, 2016).

Moreover, in Japan rapid depopulation of farming communities is taking place. The solution for this is to use farming robots to perform farm operation accurately at the right time. It allows one operator to control multiple vehicles; the goal is to get fully robotized operation in rice production. Development of three types of the robot is done. First is Robot tractor which performs tillage and puddling. Secondly, rice planter robot which performs transplanting, fertilizing and herbicide application and thirdly combines harvester robot. Robot tractor has controlled steering by stepping motor, it can be put into manual operator mode and GPS is attached to it to know its location and can be operated using computers. Many sensors are present in it for the safety of men and obstacle sensors indicates obstacles by a warning sound and warning light. This results in robot farming system into the labor of a person in Japan (Tamaki *et al.,* 2013).

Moreover, the robotic tractor and combine harvester came into existence in Japan. But for heavy weight crops existing technology is not suitable. Hence the robotic arm; end effector and control algorithm was designed. The robotic arm can be used for crop sampling and field monitoring. The harvesting robot is composed of a robot tractor, robot arm, end-effector, positioning system, vision system, and a controlling unit. The robotic arm is powered by a servo motor. It has five joints and is capable of withstanding payload of 200N with FOS- 2. Forward and Inverse kinematics are used for getting joint parameters such as joint angle, velocity and acceleration time. Using the parameters the robotic arm was designed for minimum vibration and maximum efficiency. The accuracy and repeatability are enough for the harvesting of heavy weight crops and can also be used for spraying using different end-effectors (Kamata and Roshanianfard, 2018).

We are living in a world which is ever growing. But because of this growth we face some problems also, one such kind of problem is the food scarcity. In order to stop this, we have to find new kind of techniques to increase food production. Researchers are always on this work. Now Japan is trying to overcome this problem by using the multiple robot tractors. This process consists of many stages.The laboratory of vehicle robotics has developed an electronic robot and two robot tractors. These can be very useful as they can function on their own like going forward, backward, speed steering, PTO (Paid Time Off) *etc.* They use many GPS trackers like hemisphere GPS trackers and AGI-3 GPS trackers to choose the path which they should follow. Next the safety system of sensors, here the sensors like two-dimensional laser scanners and many more are used to find out the obstacles present in their way and to change their movement. This kind of system reduces the human effort in the field setting and they can concentrate on improving the production in other ways (Noguchi, 2011).

Thereafter, many researchers are trying to introduce the robotics or mechatronics in the field of agriculture. Such kind of thing is the agricultural mobile robot. This mobile robot is designed to sense the agronomic parameters or the importance of the most important crop kinds. When designing these kinds of robots we have to make it for a universal approach to handle the crops even from one foot to even six-foot ones like jute also. In order to make this we have to take care of the diameter and frame width of the working tool primarily. In the traction, it is the system which is commonly used to distribute the loads on the soil in order to reduce the compacting and disturbing of it. This can be mainly seen in the wheels and tracks. Next comes the steering part. This is one of the important components, as it is necessary to manage the rover's direction of movement. There are many types in this like differential steering, articulated steering, and wheel steering. Next the motors and power supply which are the important components to run the robot. Then next the electronic control. Next is the mechanical structure which is the most important one as it should be as friendly as possible so that agriculture should not be disturbed (Tabile and Godoy, 2010).

Moreover, to develop the field of agriculture we have to work efficiently. People are trying to implement many changes in it. One such effort is field phenotyping robots. To construct any kind of robots, the structure is the basic designing thing. Here also structure is very important as it should work in the fields without disturbing the existing crops. It is a system which moves in a beam kind of support with a clear vision system. It clearly identifies the spot of planting and it will do the planting process with precision. The workflow of the system should be very important here. It should know the task to be performed. It should know the exact place and position of work. The height of working manipulator should be proper. As this is mostly used in the rice fields it should clearly understand the rice separating process and act on it properly. The other parameters which are needed to be taken care of are the length of arms, the height of equipment's (Zhang, 2016).

One such thing is the pollination of seeds. When insects go on the flowers for their food or the energy source, the seeds which will be present in the flower will get attached to the insects then it travels and sits on the other flower of having opposite gamete followed by fertilization. This process is called pollination. But people are trying to incorporate the robotic technology into this field. Introducing the robotic bees is not the best solution we have right now. The using of the robotic things into this field can result in an economic loss as these things cannot be done by one single robot and we require n number of them. Each one will cost a huge value. One small mistake done by the robots can result in the unacceptable high environmental cost. The ecosystem can also be damaged by this kind of things as they are made of non-degradable or in other sense they cannot be handled by nature to get things done right. By this, it can also affect biodiversity in many ways. These can not only affect the nature, but it can also affect humans by causing the food insecurity (Potts and Neumann, 2018).

The traditional farming process based up on humans. From the starting of the farming process to end process needs the human-like Ploughing, Spray seeds, Fertilizer, check the water level and availability and Harvesting. In smart farming, sensors are applied for gathering the data from various fields like Temperature sensor, Humidity sensor, Soil moisture sensor, Think to speak IoT app, Water motor, and Sprinklers. These sensors are well connected with the microcontroller. Microcontroller are received the data from the sensors and check the threshold values. Then according to the threshold values, it takes the needful action and also informed the farmer by the message. All the process completed by the Wi-Fi module. Nowadays, the culture of the poly house increasing day by day because in the poly house setup we can control the temperature, humidity, pest attack, and irrigation process. By smart farming, we can also minimize the human efforts to maximize productivity (Sneha *et al.*, 2019 and Kamelia *et al.*, 2018).

The world population is growing steadily, and now it has reached 7.7 billion people. One question that comes to mind is, what are all these people going to eat? First and foremost, this question is addressed to the agriculture industry. The growing population is hardly the only challenge facing modern farmers. What about labor shortages and consumers asking for eco-friendly sustenance? The answer to all these questions is is change towards automation and robotics in agriculture.

Chapter 2

Definition of Automation

The technique, method, or system of operating or controlling a process by highly automatic means, as by electronic devices, reducing human intervention to a minimum. A mechanical device, operated electronically, that functions automatically, without continuous input from an operator (https://www.dictionary.com › browse › automation)

We define automation as "the creation and application of technology to monitor and control the production and delivery of products and services." (https://www.isa.org › about-isa)

The technique of making an apparatus, a process, or a system operate automatically the state of being operated automatically controlled operation of an apparatus, process, or system by mechanical or electronic devices that take the place of human labor ((https://www.merriam-webster.com › dictionary › autom.)

Automation is the creation and application of technologies to produce and deliver goods and services with minimal human intervention. The implementation of automation technologies, techniques and processes improve the efficiency, reliability, and/or speed of many tasks that were previously performed by humans (https://www.techopedia.com ›. › Automation)

The use of machines and computers that can operate without needing human control office/factory automation. Automation and robotics have decreased the need for a large, highly skilled work force (https://dictionary.cambridge.org › dictionary › automati.)

Automation is a term for technology applications where human input is minimized. This includes business process automation (BPA), IT automation, personal applications such as home automation and more (*https://www.ibm.com › topics › automation)*

Automation refers to the state of being automatically operated equipments or machines and is the step beyond mechanization where it requires physical labor to perform the functions manually. With the automating the equipments, manual labor is reduced by using computer aided and automated machinery. *Automation system operation* is integrated with the use of sensory systems, feedback control loops and automatic actuating devices (*https://www.elprocus.com ›different-types-of-automati.)*

Chapter 3

Definition of Farm Automation

Farm automation (or smart farming) is a variety of tech innovations in traditional farming to optimize the food production process and improve quality. As of now, advanced farming technology can be an essential part of the farmer's daily work (https://easternpeak.com › blog › smart-farming-how-auto.)

Agricultural automation is the core technology for computer -aided agricultural production management and implementation. Specific agricultural production systems, including those related to field crops, cotton, orchards and vineyards, and animal housing and production.

Farm automation, often associated with "smart farming", is technology that makes farms more efficient and automates the crop or livestock production cycle. An increasing number of companies are working on robotics innovation to develop drones, autonomous tractors, robotic harvesters, automatic watering, and seeding robots. Although these technologies are fairly new, the industry has seen an increasing number of traditional agriculture companies adopt farm automation into their processes (https://www.progressiveautomations.com › How To 24-Apr-2020).

The heart of farm automation is the ability to track the farm field-by-field, even row-by-row. "Farm automation as we know it today was enabled by GPS technology. Before GPS, farmers managed on a field-by-field basis. Once armed with GPS, farmers could make decisions on a smaller scale, moving into precision farming."

Chapter 4

Robotics Definition

Robotics is an interdisciplinary field that integrates computer science and engineering. (German National Library". *International classification system of the German National Library (GNL).*

It means design, construction, and use of machines (robots) to perform tasks done traditionally by human beings. Robots are widely used in such industries as automobile manufacture to perform simple repetitive tasks and in industries where work must be performed in environments hazardous to humans. (*https://www.britannica.com* › robot-technology)

It is the technology dealing with the design, construction, and operation of robots in automation.https://www.merriam-webster.com/dictionary/robotics Robotics is the industry related to the engineering, construction and operation of robots – a broad and diverse field related to many commercial industries and consumer uses (*https://www.techopedia.com* ›. › Robotics).

Robotics is the branch of technology that deals with the design, construction, operation, and application of robots, as well as computer systems for their control, sensory feedback, and information processing. These technologies deal with automated machines that can take the place of humans in dangerous environments or manufacturing processes, or resemble humans in appearance, behavior, and/or cognition (*https://www.definitions.net* › definition › robotics).

It is the science of making and using robots (machines controlled by computers that are used to perform jobs automatically) (https://dictionary.cambridge.org › dictionary › robotics).

What is a Robot?

A robot is the product of the robotics field, where programmable machines are built that can assist humans or mimic human actions.

As technology progresses, so too does the scope of what is considered robotics. In 2005, 90 per cent of all robots could be found assembling cars in automotive factories. These robots consist mainly of mechanical arms tasked with welding or screwing on certain parts of a car. Today, we're seeing an evolved and expanded definition of robotics that includes the development, creation and use of bots that explore Earth's harshest conditions, robots that assist law-enforcement and even robots that assist in almost every facet of healthcare.

Agricultural Robots

Agricultural robotics aims to overcome the challenges imposed by population growth, accelerated urbanization, high competitiveness of high-quality products, environmental preservation and a lack of qualified labor. The main existing applications of agricultural robotic systems are for the execution of land preparation before planting, sowing, planting, plant treatment, harvesting, yield estimation and phenotyping (Ftobe, 2007).

Agricultural robots automate slow, repetitive and dull tasks for farmers, allowing them to focus more on improving overall production yields. Some of the most common robots in agriculture are used for: Harvesting and picking, Weed control, Autonomous mowing, pruning, seeding, spraying and thinning.

The agricultural robotics is the logical proliferation of the automation technology into the bio-systems such as the agriculture, the forestry, the green house, the horticulture and fisheries, It is replacing the common techniques to perform the same tasks with high efficiency. From high-tech greenhouses to cloud seeding, here's how agricultural robots are helping farmers fill labor shortages – and our supermarket shelves.

An agricultural robot is a robot deployed for agricultural purposes. The main area of application of robots in agriculture today is at the harvesting stage. Emerging applications of robots or drones in agriculture include weed control, cloud seeding, planting seeds, harvesting, environmental monitoring and soil analysis.

Agbot

An agbot, also called an agribot, is an autonomous robot used in farming to help improve efficiency and reduce reliance on manual labor. Future farms are expected to be tilled, sown, tended and harvested solely by fleets of co-operating autonomous robots called swarm robots that will weed, fertilize, and control pests and diseases, all the while collecting valuable data.

Chapter 5

Types of Automation

1. Basic Automation

Basic automation takes simple, rudimentary tasks and automates them. This level of automation is about digitizing work by using tools to streamline and centralize routine tasks, such as using a shared messaging system instead of having information in disconnected silos. Business Process Management (BPM) and Robotic Process Automation (RPA) are types of basic automation.

2. Process Automation

Process automation manages business processes for uniformity and transparency. It is typically handled by dedicated software and business apps. Using process automation can increase productivity and efficiency within your business. It can also deliver new insights into business challenges and suggest solutions. Process mining and workflow automation are types of process automation.

3. Integration Automation

Integration automation is where machines can mimic human tasks and repeat the actions once humans define the machine rules. One example is the "digital worker." In recent years, people have defined digital workers as software robots that are trained to work with humans to perform specific tasks. They have a specific set of skills, and they can be "hired" to work on teams.

4. Artificial Intelligence (AI) Automation

The most complex level of automation is artificial intelligence (AI) automation. The addition of AI means that machines can "learn" and make decisions based on past situations they have encountered and analyzed. For example, in customer service, virtual assistants powered can reduce costs while empowering both customers and human agents, creating an optimal customer service experience (https://www.ibm.com › topics › automation).

What are the Three Types of Automation?

1. Hard Automation (Fixed Automation)

This type of automation is referring to the use of specialized equipment used to automate fixed sequences of assembly and processing operations. The operations within the sequence, for hard automation, are typically simple, meaning they involve plain linear motion and rotational motion. In special cases, the two methods of motion can be combined. It is hard to accommodate changes in product design with hard automation.

2. Programmable Automation (Flexible Automation)

This type of automation is designed with the ability to change operation sequences that allows for the accommodation of varied product configurations. Operation sequences get controlled by a program that is based strictly off instructions coded into the system. New programs can be continually entered into equipment for the sake of producing new products.

3. Soft Automation (Flexible Automation)

This type of automation is considered an extension of programmable automation. Flexible automation systems are capable of creating a wide variety of parts with only limited amount of time loss during changeovers. The same can be said for production time when users are reprogramming systems and also altering their physical set up (*https://www.norwalt.com ›. › Cap Closing Machine).*

Major Types of Automation

1. Industrial Automation

Industrial automation is the process of making industrial production processes more flexible and simpler at higher efficiencies. Automation integration to the industries results in intelligent manufacturing solutions at improved product quality and productivity with reduced downtime and wastes. Industrial automation involves in usage of various control devices such as PC/PLC's, various sensors and actuators, communication buses/modules, machine drives, *HMI (Human Machine Interface) systems* and other control equipments.

Some of the types of *industrial automation* are given below.

a) Numerically Controlled Machines

These machines are of computer controlled machines which uses computers to perform the control operations by acquiring, processing, calculating and controlling the process variables. This automation is a programmed version of machine tools and also called as Computerized Numerical Controlled (CNC) Machines. These CNC machines are used in cutting and milling applications for high accuracy and accurate precision operation.

b) Computer – Aided Manufacturing (CAM)

In this, the entire manufacturing process (includes production, planning and control) is automated with the use of numerically controlled machines, *industrial robots* and other types of automation devices. These automation systems also make use of computers to plan, design and layout the various products. Examples of these automation systems are computer-aided design (CAD), computer-aided design and drafting (CADD) and computer-aided process planning (CAPP).

c) Industrial Robots

These are a type of automated machines or equipment's that can perform the different tasks for longer duration. These are mostly implemented in the areas that are highly dangerous or hazardous for humans.

d) Flexible Manufacturing Systems

This automation is of fully automated one. Starting from planning and designing process to dispatching of products, the whole system is completely integrated to be automated. This automation combines numerically controlled machines, industrial robots and other automation equipment's into one integrated system.

2. Home Automation

As the technologies are rapidly advancing that affects the human daily lives more flexible, particularly in controlling home appliances. Electrical installations are the heart of every building, therefore intelligent building control absolutely ensures the safety and efficient control and eventually saves the electric power consumption and human energy. *Home automation system* does the operations like lighting control and regulation, regulating HVAC (Heating, Ventilation and Air Conditioning) equipment's, energy and load management, security and monitoring, audio/video systems, HMI's and other control tasks.

Some of the types of home automation system are given below.

a) Power Line Home Automation System

This automation is of low cost since it uses power lines to transfer the information or data therefore it doesn't require additional cables for transferring the information. However, this system is of little bit complex one that necessitates additional converter circuits.

b) Wired Home Automation System

This type of automation uses a common communication bus to which all the equipments in homes are connected. This communication bus or cable is connected to a main controller (*programmable logic controller*) in order to acquire input signals and to send control command signals to the actuators.

c) Wireless Home Automation

This is an advanced automation technology as compared with wired automation. This automaton uses wireless technologies like RF, Zigbee, GSM, WI-Fi and Bluetooth for achieving remote control operation. This is a flexible type of automation requires negligible wires to connect different appliances in homes to the central control unit. (https://www.elprocus.com ›different-types-of-automati.)

Chapter 6

Concepts and Components of Robots

The concept of usage of robotics in agriculture has to comply *i.e.* better response unexpected and uncertain working environment, feasible to existing technology and economical viable than its alternatives. The conceptual idea to replicate or imitate the traditional agricultural practices by robotic technology had been tried in various unit operation of agriculture, yet no commercial robots are available of complex agricultural field conditions (Urrea and Munoz, 2015).

The basic component of a robotic system includes end-effectors, manipulators, sensors, control system, and power supply (Figure 1).

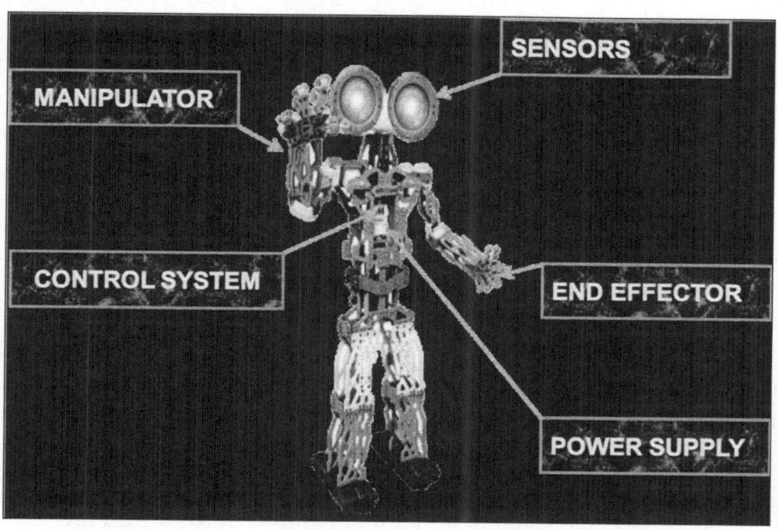

Figure 1. Components of Robot.

1. End Effectors

End effectors are the end robotic part which attach to the robotic arm or appendages, it is used for manipulating the objects by handling, griping or grasping.

2. Manipulators

They are known as robotic arm, made of finite no rigid elements called link. The links are connected to each other by joints. The commonly used joints used in robotics are revolute, cylindrical, planar, spherical, spherical, screw, and prismatic. The wrist or rotary moment is executed in x, y and z axis by roll, pitch and yaw. The three dimensional space around the robot, where it can sweep its wrist to maximum and minimum reach is called robot work volume.

3. Sensors

This are the component of the measures some aspect of the environment and convert it into a readable form. The sensors are characterised by the static and dynamic characteristics. There are two kinds of sensors used in robotics i.e tactical and wheeled sensors. Tactical sensors are used to detect the physical contact and closeness, whereas wheeled sensors are used to detect the motor position or speed.

4. Power Source

These are one of important aspect while designing the robotic system, hence care must be taken while selection of power sources and its impact on mechanism, packaging, weight and size of the system. The commonly used power sources in robotics system are generator, hybrid, batteries, photovoltaic cells and fuel cells.

5. Control System

It is used to regulate the behaviour of other subsystems; It needs to data and knowledge about current and subsequent stages of all the subsystems, which is to be controlled. The control systems are open or closed loop system. To perform a task with intended target, it is important to have an efficient control system to monitor and regulate the subsystems of robotic technology (Bikram Jyoti *et al.,* 2020).

In Agriculture autonomous robot system (ARS) are developed to perform, act and make decisions in real time unstructured and variable environment of agricultural production system. There are numerous challenges faced while designing a ARS such as real time response requirement for sensor control formulation and to model the approach to encounter the problems (Ng and Trivedi, 1998). The ARS in agriculture is comprised of various subsystems to perform different task in agricultural production system. These sub-systems and devices deal with path planning, navigation or guidance abilities, mobility, steering and control, sensing, manipulators or similar functional devices, end effectors (produce contactors or tools), and above all, guidelines on how to manage individual or simultaneous unexpected events, and some level of autonomy. Figure 2 explains the basic working system of robot in agriculture system and subsystems for its

Figure 2. Basic Working Primary and Secondary Task of Robot Agriculture.

operations. The primary task of system is to do basic agricultural operations like seeding, planting, weeding, spraying and monitoring, whereas the secondary task includes mobility, steering, navigation, manipulator and human robot interaction. The primary is designed such that they imitate the traditional unit operations of agriculture by use of end effector and manipulators. The secondary subsystem is mainly responsible for the mobility and steering sub-systems interact through soil contact, directing the vehicle in the desired orientation at the appropriate speed (Bikram Jyoti *et al.,* 2020).

Chapter 7

Types of Robots

Mechanical bots come in all shapes and sizes to efficiently carry out the task for which they are designed. All robots vary in design, functionality and degree of autonomy. From the 0.2 millimeter-long "RoboBee" to the 200 meter-long robotic shipping vessel "Vindskip," robots are emerging to carry out tasks that humans simply can't.

Generally, there are five types of robots:

1) Pre-Programmed Robots

Pre-programmed robots operate in a controlled environment where they do simple, monotonous tasks. An example of a pre-programmed robot would be a mechanical arm on an automotive assembly line. The arm serves one function – to weld a door on, to insert a certain part into the engine, *etc.* – and its job is to perform that task longer, faster and more efficiently than a human.

2) Humanoid Robots

Humanoid robots are robots that look like and/or mimic human behavior. These robots usually perform human-like activities (like running, jumping and carrying objects), and are sometimes designed to look like us, even having human faces and expressions. Two of the most prominent examples of humanoid robots are Hanson Robotics' Sophia and Boston Dynamics' Atlas.

3) Autonomous Robots

Autonomous robots operate independently of human operators. These robots are usually designed to carry out tasks in open environments that do not require human supervision. They are quite unique because they use sensors to perceive the world around them, and then employ decision-making structures (usually a computer) to take the optimal next step based on their data and mission. An example of an autonomous robot would be the Roomba vacuum cleaner, which uses sensors to roam freely throughout a home.

Examples of Autonomous Robots

1. Cleaning Bots (for example, Roomba)
2. Lawn Trimming Bots
3. Hospitality Bots
4. Autonomous Drones
5. Medical Assistant Bots

4) Teleoperated Robots

Teleoperated robots are semi-autonomous bots that use a wireless network to enable human control from a safe distance. These robots usually work in extreme geographical conditions, weather, circumstances, *etc.* Examples of teleoperated robots are the human-controlled submarines used to fix underwater pipe leaks during the BP oil spill or drones used to detect landmines on a battlefield.

5) Augmenting Robots

Augmenting robots either enhance current human capabilities or replace the capabilities a human may have lost. The field of robotics for human augmentation is a field where science fiction could become reality very soon, with bots that have the ability to redefine the definition of humanity by making humans faster and stronger. Some examples of current augmenting robots are robotic prosthetic limbs or exoskeletons used to lift hefty weights. From Diffusing Bombs to Performing Surgery, VR Robots Have Some Amazing Uses.

6) Industrial Robots

Industrial robots often comprise the most basic form of machine – a stationary or semi-stationary device that executes a repetitive task. These robots are generally some of the least intelligent due to the fact that the work they do is incredibly simple and the environments in which they work are fairly free of external influences that could disrupt their routines.

7) Exploration Robots

These cybernetic adventurers can range in complexity from simple probes to fully autonomous spacecraft. They are used to explore the farthest reaches of space and the darkest trenches of the ocean floor, boldly going where no man has gone before.

Some of the more famous examples of these, such as the Mars Rover Opportunity, are a type of robot known as Remotely Operated Vehicle (ROV) that performs some autonomous functions while having the capability of being operated by a remote operator or pilot. These robots typically come equipped with advanced observation or manipulation features. These features allow them to gather data from their assigned environments in a more focused way than their less complex aerospace cousins.

8) Consumer Robots

Consumer robots are so commonplace that many people fail to see them as robots at all! These are the little household helpers that unobtrusively improve the lives of countless homeowners the world over. The classic example of a consumer robot is the roomba, an autonomous cleaning machine complete with sensors to help it navigate any space you put it in.

More recently, simplistic AI has started to fuse with consumer robots and the internet of things to give us devices such as the Amazon Alexa or Google Home. These stationary robots come equipped with what's referred to as a 'conversational AI' that can read things such as tone and context to make an educated guess as to the intent of the speaker. While these devices are still far from perfect, it's exciting to imagine the quality of life they could provide us in the next half century.

9) Medical Robots

Aside from production, exploration, and menial tasks, robots can also be literal life savers. Medical robots can range from autonomous arms that help surgeons perform delicate operations to the emerging field of mind-controlled robotic prosthetics and exoskeletons. While we probably won't see fully autonomous surgeons for many years to come, doctor-operated robots have pushed the boundaries of medicine in terms of what can be accomplished without having to resort to risky invasive procedures. Devices such as the daVinci system have made it possible to perform dangerous operations with a fraction of the normal risk.

10) Aerospace Robots

Aerospace robots are, in some capacity, able to fly. They differ from exploration robots in that they don't include aquatic automatons or surface rovers. Common forms of these robots are autonomous or remote controlled drones or spacecraft that can be used for a variety of purposes such as research, military intelligence, or deep space exploration.

11) Aquatic Robots

Aquatic robots go far beyond deep-sea exploration. They can work with the coast guard as unmanned boats and have often been used by marine biologists and conservationists to help supplement parts of marine ecosystems that have been ravaged by climate change and industrialization efforts like off-shore drilling.

Interestingly, Aquatic robotics is also one of the subfields at the forefront of biorobotics, where the robots developed take inspiration from organic animals in the wild for how they move. Many researchers have found great success with modeling their swimming robots off of common underwater species such as fish or eels, whose innate abilities to navigate the underwater seascape are far superior to traditional methods of locomotion for human-made sea-craft (Piper Thomson, 2019).

Robotics Companies to Know

1. Nuro
2. CANVAS Technology
3. Piaggio Fast Forward
4. Diligent Robotics
5. Boston Dynamics
6. Bluefin Robotics
7. Applied Aeronautics
8. Left Hand Robotics
9. Righthand Robotics
10. Dronesense
11. Harvest Automation
12. Rethink Robotics
13. Vicarious

Chapter 8

Some of the Main Applications of Robots for Farming

Agriculture is quickly becoming an exciting high-tech industry, drawing new professionals, new companies and new investors. The technology is developing rapidly, not only advancing the production capabilities of farmers but also advancing robotics and automation technology as we know it.

At the heart of this phenomenon is the need for significantly increased production yields. The UN estimates the world population will rise from 7.3 billion today to 9.7 billion in 2050. The world will need a lot more food, and farmers will face serious pressure to keep up with demand. Agricultural robots are increasing production yields for farmers in various ways. From drones to autonomous tractors to robotic arms, the technology is being deployed in creative and innovative applications. Agricultural robots automate slow, repetitive and dull tasks for farmers, allowing them to focus more on improving overall production yields. Some of the most common robots in agriculture are used for:

1. Harvesting and picking
2. Weed control
3. Autonomous mowing, pruning, seeding, spraying and thinning
4. Phenotyping
5. Sorting and packing
6. Utility platforms

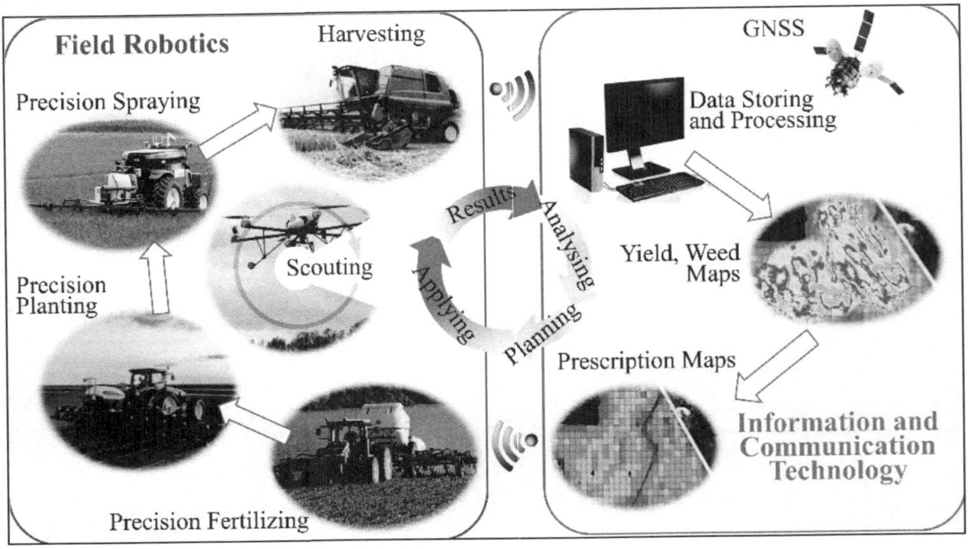

Figure 3. Field Robotics and ICT in Farming.

Harvesting and picking is one of the most popular robotic applications in agriculture due to the accuracy and speed that robots can achieve to improve the size of yields and reduce waste from crops being left in the field. These applications can be difficult to automate, however. For example, a robotic system designed to pick sweet peppers encounters many obstacles. Vision systems have to determine the location and ripeness of the pepper in harsh conditions, including the presence of dust, varying light intensity, temperature swings and movement created by the wind.

But it still takes more than advanced vision systems to pick a pepper. A robotic arm has to navigate environments with just as many obstacles to delicately grasp and place a pepper. This process is very different from picking and placing a metal part on an assembly line. The agricultural robotic arm must be flexible in a dynamic environment and accurate enough not to damage the peppers as they're being picked.

Harvesting and picking robots are becoming very popular among farmers, but there are dozens of other innovative ways the agricultural industry is deploying robotic automation to improve their production yields. The demand for food is outpacing available farmland and it's up to farmers to close this gap. Agricultural robots are helping them do just that. Agriculture has quickly become a high-tech business. If you want to learn more about emerging robotics industries, join companies like Amazon, Microsoft and GM at the annual A3 Business Forum to network with leaders in the automation industry (Qin Zhang, 2013).

As the human population grows, farmers have to use new technologies to keep up with growing demand. By 2050, there are expected to be about 9 billion people in the world. IEEE Robotics and Automation Society says that "agricultural production must double if it is to meet the increasing demands for food and bioenergy." In order to meet the 9 billion people's demand for food, robotics and automation are expected to play a major role in society.

- Equipment manufacturers are developing tractors that use machine vision and autonomous robotics to replace traditional large, expensive and inefficient tractors.
- In the rules-oriented organic farming field, robotic weeding implements have been used to inspect crop rows and identify weeds rapidly.
- Robots operating at low speeds can detect weeds. Next-generation computer vision algorithms will be used to classify them.
- Robots are being used for harvesting and sorting, packaging and boxing.
- Other future applications include planting seeds, grafting plants, seeding clouds, analyzing soil and monitoring the environment (*https://whatis.techtarget.com* › definition › agbot-agricultu.).

Robots can be used in many situations and for many purposes, but today many are used in dangerous environments (including inspection of radioactive materials, bomb detection and deactivation), manufacturing processes, or where humans cannot survive (*e.g.* in space, underwater, in high heat, and clean up and containment of hazardous materials and radiation).

1. Robots and Autonomous Machines

Agriculture is a perfect niche for innovations in the sphere of robotics: farmers usually have to deal with repetitive tasks in the field, and this work is primarily labor-intensive. Now agricultural robots (or "agrobots") cope with a wide range of tasks: harvesting, watering, seeding, *etc.* Let's take a closer look at their work.

2. Automated Tractors

Automated Tractors.

These types of tractors can be controlled remotely. It is only necessary to help during the initial set-up and maintaining machines regularly. With time, smart tractors will become more and more independent with such technologies as vision systems, tools for light detection, GPS, *etc.* Bear Flag Robotics specializes in building driverless tractors and aims to make the labor cheaper for farmers. They can control several tractors remotely, plan the machine's route, get real-time reports and alerts – the work can be done without direct human intervention in the field.

3 Seeding and Weeding Robotics

Robots for planting are focused on the field's specific area, and they work with great precision. This type of farming robot uses artificial intelligence and computer vision, which allows for a reduction of pesticides in the field and, subsequently, the production of high-quality food. Eco Robotics produces completely autonomous robots that use solar power for their work. This is the greenest approach: a small four-wheel machine is moving across the field and spraying herbicides with minimum damage to the crops and environment.

Weeding Robotics.

4 Automatic Irrigation

Robot-assisted irrigation systems roughly contain two big parts: a Subsurface Drip Irrigation (SDI) system plus special sensors. SDI is well-known in the agriculture industry – it provides an accurate means of controlling the amount of water that is used, and at what time the plants receive it. Even though these systems are obviously more advanced than manual plant-by-plant watering, they are still not perfect as they require some human assistance. More sophisticated IoT sensors can monitor the moisture levels by themselves and send real-time analytics to a smart device. The combination of SDI and such sensors creates automated farming equipment that facilitates the work and saves water resources. The RAPID (Robot-Assisted Precision Irrigation Delivery) project offers farmers a cost-conscious solution based on the work of adjustable irrigation emitters across the field.

RAPID Robos.

5. Drones

Drones have an important role in agriculture's innovative changes. Earlier, having a glance at a field was possible only with helicopters or even satellites – things have changed now. Using a drone is cheaper and does not require any special human skills like flying a helicopter. Drones can take aerial photos and videos, and the files' quality is getting better with time. But this is far from the only way to use drones in agriculture. The activities of drones such as:

a) Monitoring and Analytics

The average field size makes it almost impossible to properly monitor the state of a harvest. Previously, the most reliable solution was satellite imagery, but still, it

Drone Flies over Field and Capture Images of Crop Condition.

did not give enough guarantees to farmers: such images were inaccurate and did not provide real-time information. With drones, automation in farming is brought to the next level. Drone devices can identify damaged crops using different types of light, including near-infrared. Farmers receive this information immediately and can quickly react to the changing conditions (check the eBee SQ drone developed by senseFly). Drones are also useful when creating a planting scheme. They provide 3D field maps with soil analysis – it helps farmers make the right decisions at the very beginning of the crop cycle.

Figure 4(a-f). Unit Operation-wise Application of Robotics in Agricultural Production System.

Figure 4(g-l). Unit Operation-wise Application of Robotics in Agricultural Production System.

b) Planting

Automated farms can use robots for seeding, not only on the ground, but also for planting from the air. Let's see how it works. The main goal of DroneSeed is to help reforest lands after wildfires. The drone carries a module that "shoots" seeds at the set place and specific distance from each other. To make it possible, developers use compressed air – this way, seeds will not scatter around the field. Forestry, however, is not the only branch which requires the help of a drone. We might see "air planting" used more often as a part of agriculture automation.

c) Irrigation

And finally, you can use drones to water plants in the field. Thanks to the latest innovations, such drones adapt to several location factors: altitude, objects in the field, weather conditions, *etc.* Agras MG-1's drone created by DJI is powerful (it can carry up to 10 liters of liquid) and fast (it covers up to 6 000 m^2 in 10 minutes). It saves lots of time and effort – manual watering is dozens of times slower. If several farming automation solutions work in conjunction, the watering drones can get signals from other agrobots about the plants that need the help more than others. In other words, agrobots communicate with each other and solve problems on the go.

6. Weeding Robots

Various researchers had worked to develop robotic system in agriculture. Nishiura *et al.,* 2019 fabricated robot that could work as 8 persons when the improved soft handling and alignment were carried out. The system was developed for raising and grafting of seedling of y, Y and V type with success rate of 95, 93.8 and 46.8 per cent respectively, Figure 4a. An autonomous RTK GPS guided weed-control system to remove or destroy weed. An intelligent hoe uses vision systems to identify the rows of crops, and steer itself accurately between them, considerably reducing the need for herbicides, Figure 4b (Slaughter *et al.,* 2008).

EcoRobotix's fully automatic robot was designed to destroy weeds. The system has an efficiency of more than 95 per cent to spray at right spot with minimum damage. The robot can be modified to adapt to new or changing environment variables Figure 4c (EcoRobotix, 2019). An innovative agricultural robot for crop and weed management, intended to work in for horticultural crop management, Figure 4d. A Robotic in-row weed control in vegetables was developed with DoD-system to adjust differences in cultivation methods, number of crop rows, track width and height of the crop. The system reduced the manual drudgery 50-100 per cent in a grown sugar beets and vegetables and herbicide usage by 75- 100 per cent in high value crops, Figure 4e (Utstumo *et al.,* 2018). The robotic weeding was developed for weed control in maize. The weeder has RKT DGPS to guide the tractor during working, Figure 3e various researchers globally has globally tried robotic system for tomato harvesting, weed management, spraying using semiautomatic teleoperation of an agricultural robotic system to enable improved performance in Figures 4f,g,h,i,j,k and l (Tijmen *et al.,* 2010, Hong *et al.,* 2009, Adamides *et al.,* 2014, Oberti *et al.,* 2013 and Sammons *et al.,* 2005).

Chapter 9

Robotic Applications in Agricultural Operations

Indian researchers have also tried to develop various kind of robots for agricultural operations. TNAU has developed a pneumatic actuated gripper for grasping and releasing the plug type seedling. The transplanting rate of developed mechanism was set as 20 – 25 seedlings/minute, Figure 5a. TNAU has also developed a robotic rice pellet seeder. The control of the seeder is done remotely from the telemetry controller. The basic features of the controller are the two Joy sticks for right and left moment 5b. IARI New Delhi has developed a robotic precision planter with wireless control through microprocessor using Wi-Fi module. There are Traction wheel controls, steering control, seeding mechanism control in three Cartesian coordinates. The total time for completion of one block is 3 min 40s, Figure 5c. A commercial robot named GRobomac has been developed by a Bangalore based robot firm for detection and harvesting of cotton balls, Figure 5d. IIT Kharagpur has developed a robotic arm for cotton picking based on 3D machine vision techniques, Figures 5 e, f.

1. Land Preparation before Planting

Preparing the land before planting is one of the first agricultural tasks to be carried out, such as plowing the land and applying fertilizers. The practice of plowing the land (inversion of soil layers) allows a greater introduction of oxygen and expulsion of carbon dioxide; however, depending on the local climate conditions, it can harm future crops by considerably reducing the soil's carbon stocks. On the other hand, soil fertilization serves to replace the nutrients necessary for the development of crops. The review article, published in 1987, refers that one of the main challenges in the development of robots that operate in rough terrain, as in a plowed field, is the creation of a robotic system that is precisely controlled (Sistler, 1987).

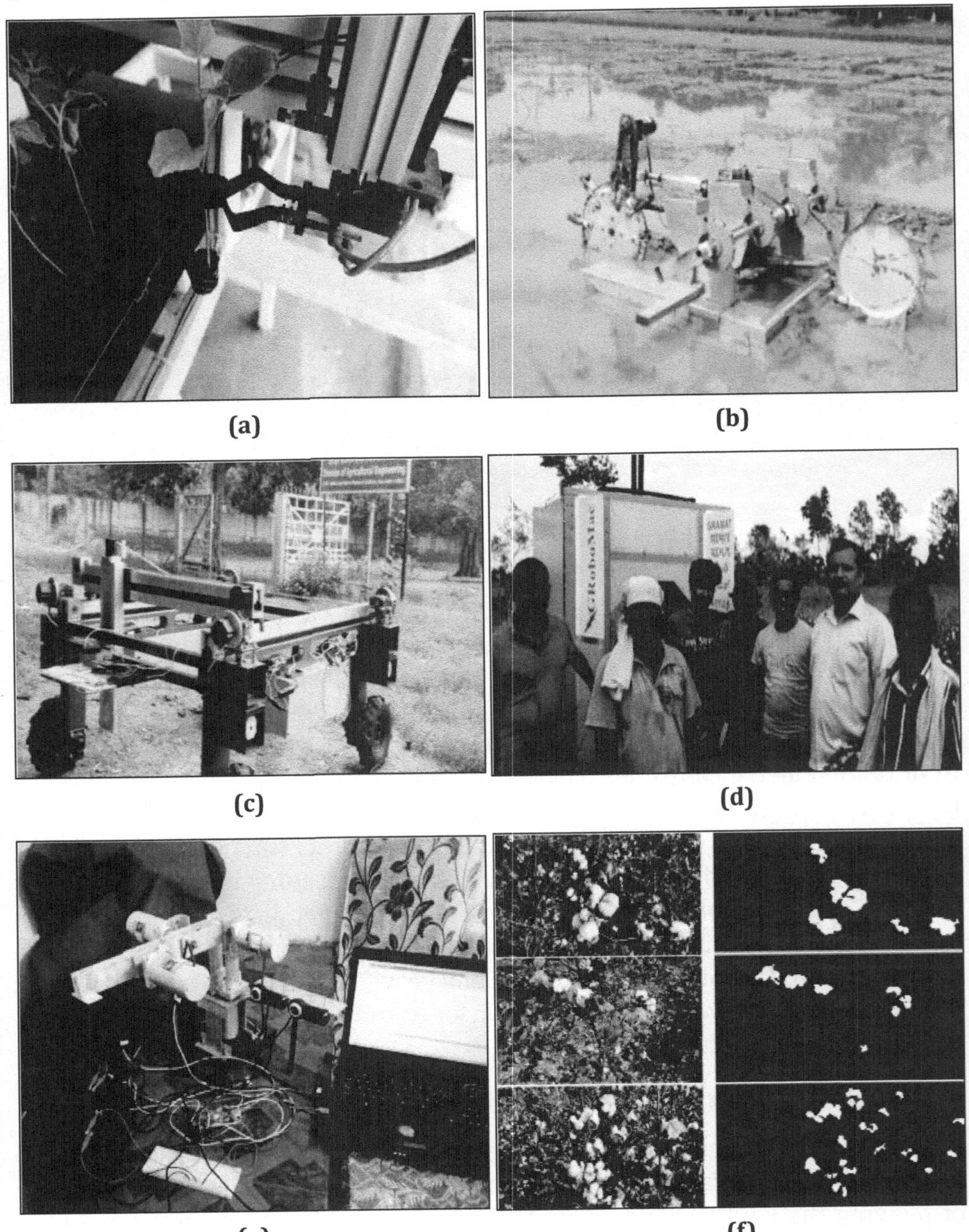

Figure 5. Robotic Application in Indian Agriculture.

To create a robot capable of assisting rural workers not only in soil fertilization but also for pest control, soil management, harvesting and transportation, the German company Raussendorf developed in 2014, the robot entitled Cäsar. The commercially available Cäsar robot can fertilize the soil in two ways: via remote control or autonomously (Siciliano and Khatib, 2016).

To perform tasks automatically, it uses Real-Time Kinematic (RTK) technology for the Global Navigation Satellite System (GNSS). GNSS is the navigation device, however, it can use different services, such as GPS (North American), GLONASS (Russian) or GALILEO (European), resulting in a location accuracy of up to 3 cm. Designed to work on the farm together with human workers, the Cäsar robot has a collision detection system based on ultrasonic sensors that guarantee its immediate stop, with a maximum detection distance of 5 m (Khan *et al.*, 2018 and Nawaz *et al.*, 2012).

2. Sowing and Planting

Traditionally, sowing and planting tasks are performed by specialized planting equipment connected behind the tractor. However, tractors are characterized by being heavy machinery and, therefore, their constant locomotion throughout the farm intensifies soil compaction. The soil compaction activity has several damaging effects for the agricultural environments, such as increasing the apparent density and soil resistance, reducing porosity, the rate of water infiltration and aeration, in addition to affecting the chemical properties and biogeochemical cycles, affecting the development of plants and soil biodiversity (Sakaue, 1996).

In 1996, Sakaue was already developing robotic systems to automate the sowing and planting process in Japan with the elaboration of four variations of a wheeled robot. Despite having a rudimentary structure, prototype number four was capable of planting cabbage, lettuce, broccoli, cauliflower or celery of up to 2200 plants/h. However, the system was not able to move autonomously (Haibo *et al.*, 2015).

Motivated to produce a small, high precision and capable of agile and easily moving along Chinese wheat farms, Haibo *et al.* (2015) developed the Lumai-5 robot. For this type of task, the robot must ensure that the sowing process remains constant, even for different speeds of detachment. The Lumai-5 robot has a 4WS, a closed-loop control system and speed, angle and pressure sensors to accurately carry out wheat sowing. The size of the planting tray, the vacuum chamber pressure and the planting speed were the main factors that directly affected the seeding quality (FAO, 2021).

3. Plant Treatment

After finishing the sowing stage, maintaining the planting growth properly, that is, free of diseases and pests, requires constant monitoring by the farmer. According to FAO data, about 20 to 40 per cent of world crop production is lost due to pests and diseases (Sinden, 2004).

The good growth of crops invaded by weeds is strongly compromised and can even destroy crops completely. Weeds can also attract pests in addition to small animals like snakes and mice. Thus, the sooner it is extracted, the lower the financial losses. As an example of these costs, in Australia weed management costs are close to $4 billion a year (Lee *et al.*, 1999).

Plant treatment is commonly done by applying herbicides and pesticides (insecticides and fungicides). The searches for the automation of the diseases identification process in plants and detection of weeds are not recent–there have already been works in this area since 1998. In a robotic weed control system for tomatoes is described, based on the Bayesian classifier algorithm that correctly identified 73.1 per cent of tomatoes and 68.8 per cent of weeds in the validation set of field images. In researchers Lee and Slaughter (1998) chose to develop the hardware-based neural network, rather than the Bayesian classifier, for increasing the plant identification. With this new classification method, the robotic system correctly identified 38.9 per cent of tomato cotyledons and 85.7 per cent of weeds. These were just the first robotic systems applied to weed control, from a series of robots that have been developed over the years. Therefore, this subsection will address several works related to these agricultural tasks, focusing on its latest developments.

Using a 6 Degrees of Freedom (DoF) manipulator arm, an RGB camera and a laser distance sensor (DT35, SICK), all mounted on a fixed platform, Schor *et al.* (2016) developed a robotic system to perform the detection of powdery mildew and tomato spotted wilt virus diseases in greenhouses. The RGB camera and the laser sensor were placed on the final actuator of the manipulator to extract the images from different angles and avoid collisions with the plant. The images were used in the disease detection process, using Principal Component Analysis (PCA) and Coefficient of Variation (CV). The system obtained an accuracy rate of 64 per cent for the process of classification of plants with powdery mildew disease in an early stage of evolution and up to 90 per cent for the case of tomato spotted wilt virus, thus enabling accurate detection of diseases in its initial stage.

For the same purpose, the mobile robot eAGROBOT was used to identify pests in cotton and groundnut crops. Through the application of artificial intelligence algorithms, such as artificial neural networks and K-means, in the images acquired by an RGB camera of crops in the initial sowing stage (period of the emergence of diseases such as leaf spot and anthracnose), the robot achieved a precision of 83–96 per cent for disease identification in normal images and 89 per cent for wide images (Pilli *et al.*, 2015).

4. Harvesting

In addition to being a repetitive task and requiring agile execution, the harvesting activity requires a lot of effort from the harvester. According to Hayashi *et al.* (2014), harvesting operations represent around 25 per cent of all hours of agricultural work in Japan.

Concerning financial costs (an important decision factor in farmers' decision-making) in 2014, in Australia, labor rent represented between 20 per cent and 30 per cent of total cash costs and in China, in 2019, it represented more than 75 per cent of production costs and increasing annually. This difference in values, between Australia and China, is due to the higher level of automation of the tasks of Australian crops. In this way, several works are being carried out proposing the use of robotic systems to perform agricultural harvesting activities. Several scientific articles address different image processing techniques in different types of cultures (Jie, *et al.*, 2019).

During the design and implementation of an aided fruit-harvesting robot, Ceres *et al.* (1998) found that the main difficulties in the development of these types of systems were: driving the robot through the field from tree to tree and from tree row to tree row; detection and localization of fruits, grasping and detaching of selected targets. Another relevant work carried out in the 1990s was the autonomous mobile robot AURORA for greenhouse operation, capable of autonomously navigating the greenhouse corridors using only the information provided by ultrasonic sensors. The robot was built to be a multitasking platform, to perform tasks such as: harvesting, transportation of fruits and plan inspection. In this case, the project requirements were navigation in unaltered greenhouses, low cost, flexibility, multi-functionality, supervizable autonomous operation and friendly user interface.

5. Yield Estimation and Phenotyping

Through accurate data on the quantity and quality of the fruits growth, provided by more sophisticated tools, farmers can manage their crops more efficiently. Yield estimation is nothing more than monitoring the entire crop and estimates the produced fruit. On the other hand, several variables such as climate change and soil quality can interfere with the development of plants. Thus, by identifying the phenotype of the plants, it is possible to link to their respective genotype, making it possible to identify the proper growth conditions. It is noted, however, that for a robot to estimate phenotyping and/or yield it must have not only reliable sensory data, but also efficient computer vision algorithms. In 1998 and 2001 there were already researchers proposing the use of sensors and machine vision algorithms to detect crop rows and gather field information (Noguchi *et al.*, 1998 and Noguchi *et al.*, 2001). In this case, a camera and a RTK/GPS device were inserted in a tractor to be able to create spatial maps relating the crop height and width. Through ANN the robotic system obtained a hit rate of 84 per cent, indicating that a machine vision system could be used as a crop prediction sensor. Thus, this subsection will address several robots developed over the years to perform the tasks of yield estimation and phenotyping, giving an emphasis on the ones developed in recent years. The Shrimp robotic system equipped with six RGB cameras was used to estimate the yield of apple orchards under natural lighting conditions. To locate each sampled image, the Shrimp platform relies on the integration of an Inertial Navigation

System (INS) with a GPS. The image processing is based on Multiscale Multilayer Perceptron (MLP) and CNN and the detection of the apples was performed through Watershed (WS) segmentation and Circular Hough Transform (CHT). The Shrimp platform obtained an apple detection of 82.5 per cent, F_1 of 0.791 and a coefficient of determination r^2 of 0.826, using CNN and WS (Bargoti and Underwood, 2017).

Chapter 10

Top 10 Robotic Applications for the Farmers

1. Nursery Planting

Nurseries are where seeds are grown into young plants, which are later planted outside. Nursery plants are often sold direct to consumers and landscape gardeners, but they are also the start of the food journey for some crops. There is a rising need for nursery automation. Companies like HETO Agrotechnics and Harvest Automation solutions for seeding, potting and warehousing living plants in greenhouses.

2. Crop Seeding

Many food plants begin life as seeds in a field. The traditional method for sewing seeds is to scatter them using a "broadcast spreader" attached to a tractor.

Image Courtesy Crop Seeding Robos.

This throws many seeds around the field while the tractor drives at a steady pace. It is not a very efficient method of planting as it can waste seeds.

Autonomous precision seeding combines robotics with geomapping. A map is generated which shows the soil properties (quality, density, *etc.*) at every point in the field. The tractor, with robotic seeding attachment, then places the seeds at precise locations and depths so that each has the best chance of growing.

3. Crop Monitoring and Analysis

Monitoring huge fields of crop is a big job. New sensor and geomapping technologies are allowing farmers to get a much higher level of data about their crops than they have in the past. Ground robots and drones provide a way to collect this data autonomously. Drone companies like Precision Hawk offer farmers combined packages which include robotic hardware and analysis software. The farmer can then move the drone to the field, initiate the software via a tablet or smartphone, and view the collected crop data in real time. Ground based robots, like BoniRob, provide even more detailed monitoring as they are able to get closer to the crops. Some can also be used for other tasks like weeding and fertilizing.

4. Fertilizing and Irrigation

Irrigating and fertilizing crops has traditionally used a lot of water is quite inefficient. Robot-Assisted Precision Irrigation can reduce wasted water by targeting specific plants. Ground robots autonomously navigate between rows of crop and pour water directly at the base of each plant.

Robots also have an advantage as they are able to access areas where other machines cannot. For example, corn growers face a problem that the plants grow too quickly to reliably fertilize them. Rowbot aims to solve this problem as it easily drives between the rows of corn and targets nitrogen fertilizer directly at the base of each plant.

5. Crop Weeding and Spraying

Spraying pesticides and weed killers onto fields is not only wasteful and it can severely harm the environment. Robots provide a much more efficient method. The concept of micro-spraying could significantly reduce the amount of herbicide used in crop growing. Micro-spraying robots use computer vision technology to detect weeds and then spray a targeted drop of herbicide onto them. AG BOT II is a solar powered robot which uses this technique. Some weeding robots don't even need to use chemicals. RoboCrop, for example, uses computer vision to detect plants as it is pushed by a tractor. It then automatically hoes the spaces between plants to uproot the weeds. Other weeding robots use lasers to kill the weeds.

6. Thinning and Pruning

Thinning involves reducing the density of plants so that each has a better chance of growing. Pruning involves cutting back parts of plants to improve their growth. The LettuceBot thinning robot received an award this year for "outstanding product innovation in agriculture." It uses computer vision to detect lettuce plants as it drives over them and decides in that moment which plants to keep and which to remove. Pruning is a complex job and the most notable attempts to automate it have come in the wine industry. Wall-Ye is an autonomous vineyard robot able to prune grape vines. The company has also developed a blueberry pruning robot.

7. Autonomous Tractors

Several of the robots mentioned are attachments to a tractor. As humans usually drive the tractors, the robots are designed to adapt to the speed that the human is driving. However, fully-autonomous tractors are also becoming popular.

Autonomous Tractors.

The IDTechEx Agricultural Robots report found that more than 300 thousand tractors with autonomous functionality were sold in 2016. There is also a rising trend for follow-the-leader autonomy, where tractors autonomously follow human-driven combine harvesters to collect the grain.

8. Picking and Harvesting

Harvesting crops such as corn, barley and wheat is quite simple. It can be done with combine harvester, which can be automated just like a tractor. However, other crops, like soft fruits, are more difficult to harvest as they require manual dexterity. The EU-funded "Clever Robots for Crops" project is making progress on a few harvesting applications, including apple harvesting, grape picking and sweet pepper picking.

9. Shepherding and Herding

Although most agricultural robots are currently applied in crop growing, there have been a few emerging applications within sheep and cattle farming. Farmers in both New Zealand and Ireland have been seen using drones to round up their sheep over long, difficult terrain. Various examples of remote-controlled, cattle-herding drones can also be found on YouTube. This could be a lower cost alternative to the expensive helicopters which Australian ranchers use currently.

10. Milking

Finally, we have seen a great development using collaborative robots to help in the milking process on dairy farms. This case study by Universal Robots shows how a UR5 can be used to spray disinfectant on the cow's udders in preparation for milking.

UR5 Robots.

It seems certain that robotics will continue to revolutionize agriculture and change the way that we think about producing food. For a list of even more agricultural robots, check out these articles at Robotics Business Review, The Robot Report and IntoRobotics (Alex Owen-Hill, 2017).

Chapter 11

IoT in Agricultural Automation

In the world of today, automation reigns supreme. It is leveraged in a plethora of spheres – from office furniture and domestic environment mechanization to medicine and various industries. Agriculture is no exception where modern farming tools and agricultural automation can help to streamline crop production cycle making it more efficient and less labor-consuming.

This vector towards farming automation has become especially apparent in the new millennium with more money channeled into agricultural startups and automated farm systems. TechCrunch estimated the capital investment in such companies at an astonishing figure of $1.5 billion in 2017 which is seven and a half times more than ten years before. The number of startups manifests a significant growth as well with 160 agriculture automation companies vying for financing in comparison to 31 in 2007.

The list of modern farming techniques encompasses all aspects of crop production. Some companies introduce software allowing efficient seed and fertilizer management as well as soil and irrigation control down to the prediction of yield. Other startups offer drones that are leveraged not only with surveillance purposes but can apply crop treatments from above such as fertilizers, pesticides, herbicides, and the like. Still others design and employ automated farm equipment with a range of applications – from grain augers, combines, and automated tractors to fruit picking robots. However, beckoning such prospects may seem farmers realize that automated agriculture starts not with modern farming equipment but with proper intelligence and reconnaissance that is data collecting.

The accurate information about the condition of your farm is delivered by sensors. For instance, remote sensors manufactured by CropX are placed all over the field to notify farmers about an inadequate state of the soil like the lack of moisture in it. However, automation in agriculture makes the immediate feedback of the farmer redundant since automated farming systems include other devices connected to the sensors and gauges to react to the obtained data following the preset steps and procedures. So the irrigation system switches on conveying water when and where it is most wanted.

Such modern farming technology is called the IoT and it doesn't require the presence of a human to control various agricultural processes. Instead, sensors, gauges and machines function on their own being linked together via the cellular network, mobile Internet, Bluetooth, or any other type of wireless connection. Application of IoT at an automated farm provides resource delivery optimization and maximum efficiency of precision agriculture both of which spell greater yields of higher-quality crops. The automated farming equipment connected through the Internet is expected to manifest a 20 per cent growth over the next few years which are explained by its potential to be employed at each step in the agriculture value chain.

Managing a farm is a complex activity involving dozens of operations the implementation of which is conditioned by numerous factors. These factors influence the farmers' decisions as to the choice and application of seeds, fertilizers, pesticides, and contain information about the weather conditions, expected rainfall, air, and soil temperature, *etc.* This data is provided by sensors that trigger certain software algorithms that optimize corresponding agricultural processes at automated farms. Another potential domain for agricultural IoT application is related to resource tracking and management. By employing respective software farmers will be able to register all field applications with the subsequent tracking of these resources to storage locations and elevators. As a result of such optimization, food, and beverage manufacturers can excel at creating marketing value for their businesses.

Modern farming technologies greatly benefit from employing IoT in storage safety. For example, equipping silos and elevators with OPI Systems' sensors enables to monitor the inside environment conditions and warn people about excessive heat or moisture that can be detrimental to the stored grain or send alerts if a fire is imminent. Automation in farming isn't limited to the ushering of state-of-the-art technologies. It encompasses upgrading conventional and introducing modern agricultural tools and machines.

1. Modern Farm Machinery

An automatic farm boils down to the participation of as little human element as possible which presupposes leveraging self-controlled machines and robots. What impedes a large-scale introduction of such autonomous modern farming equipment is its high price and restricted commercial availability. However, the contemporary mechanization trend spurs many manufacturers to develop cost-effective and reliable farm automation machines.

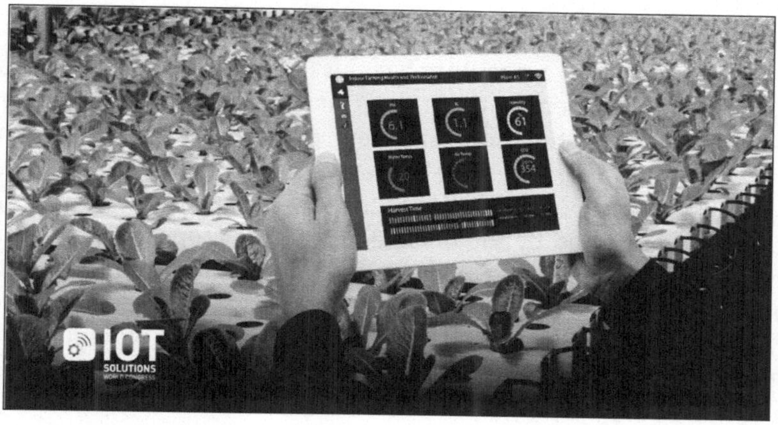

Modern Farm Machinery.

What is modern farming automated machinery as most of us imagine it? It is probably the mental picture of a tractor, combine, or motorized auger that does all the work by itself without any human participation or indeed intervention. Sharing this vision many companies try to implement it developing their new models with the increasing employment of autonomous features. For instance, modern tractors by John Deere have fully automated line keeping and depth adjustment functions.

Other major tractor manufacturers move along similar tracks aspiring to design remotely controlled and pre-programmed machines that would allow their users to cut down on labor and input costs. Some of their projects have already come to fruition. DOT Technology Corp produced a diesel-powered seeder that can sow an entire field with no human interference by following predefined routes. Rowbot has come up with a modern agriculture machine that can spread nitrogen between corn rows and sow cover crops before the ripe corn is harvested late in the season. To crown all such disparate agriculture automation endeavors, The Hands-Free Hectare project was launched in the UK where completely autonomous farming machines planted, cultivated, and harvested a hectare of barley with no human setting a foot into the field.

Having made an automated wheat farm a reality of the nearest future manufacturers are setting daring plans in other agricultural spheres including those which have traditionally been considered hard to automate. One such domain is horticulture where fruit and vegetable harvesting is a delicate procedure since it should prevent both bruising of gentle crops and damaging plants. First essays in this realm were made by Abundant Robotics and Energid that built robots for picking apples and oranges respectively. They are just pilot samples that are to yet be developed into viable models susceptible to employment in the real orchard conditions.

2. Elevated Farming

A modern machine in agriculture isn't just a ground crawler like a tractor or a grain auger. Some of them watch crops from on high providing information of their state as well as identify problem zones, spray chemicals or even blow water off cherry trees ready for harvesting to prevent berries from bursting after heavy rainfall. The use of drones is especially beneficial in cramped row spacing conditions which can be exacerbated by the hillside location of crops. Such conditions are in evidence at Napa vineyards which makes fungicide application a particularly taxing activity. So instead of farmworkers obliged to carry heavy sprayers on their backs, the task was entrusted to a remote-piloted helicopter by Yamaha Precision Agriculture which significantly streamlined the procedure sparing humans a literally backbreaking work.

Elevated Farming.

With the introduction of automation, agriculture becomes increasingly user-friendly with farmers spending less time in the field and more time at computers analyzing data and diagnosing problems. So far completely automatic farms are still a coveted dream but rapid development and sophistication of agricultural machines pave way for the advent of automatic farming as a mainsteam trend of the modern evonomy. (*https://www.progressiveautomations.com › How To 24-Apr-2020*)

Chapter 12

Automation and Wireless System Networks in Agriculture

Agriculture automation is the main concern and emerging subject for every country. The world population is increasing at a very fast rate and with increase in population the need for food increases briskly. Traditional methods used by farmers aren't sufficient enough to serve the increasing demand and so they have to hamper the soil by using harmful pesticides in an intensified manner. This affects the agricultural practice a lot and in the end the land remains barren with no fertility. There are some areas which are causing the problems to agriculture field like crop diseases, lack of storage management, pesticide control, weed management, lack of irrigation and water management and all this problems can be solved by automation and wireless network. Today, there is an urgent need to decipher the issues like use of harmful pesticides, controlled irrigation, control on pollution and effects of environment in agricultural practice. Automation of farming practices has proved to increase the gain from the soil and also has strengthened the soil fertility (Elsevier, 2019).

1. Technology Roadmap (TRM)

It is imperative for any sector to evolve with time. The agriculture sector had to adapt the breakthroughs and inventions which came along in automation field. Yong *et al.* (2018) came forward with emerging research area of embedded intelligence (EI). Embedded intelligence in agriculture sector includes smart farming, smart crop management, smart irrigation and smart greenhouses. It is necessary for a nation to include these growing technologies in agriculture sector for growth of a nation as many sectors are inter-dependent on agriculture. Also, researchers of this paper demonstrated TRM which in turn clarifies the qualms regarding the areas of agriculture mentioned above (smart farming, smart irrigation *etc.*). Use of wireless communication has changed the standards of communication in today's world and this can also raise the standards of agriculture automation. Logatchevl *et al.* (1998) has bifurcated the IOT gateway into different nodes such as actuator, sensor, interface and wireless link which give assistance to communication between them. Frequency estimation and the bandwidth requirement for this communication has also been elaborated which can be very beneficial for automation.

2. Predicted Grape Disease

Taking into consideration the socio and economic vitality of agriculture in India, researchers Patil and Thorat (2016) developed a system which predicted grape disease beforehand. Any anomaly in the grape plant was noticed only after it was infected and this had a considerable deteriorating effect on the whole vineyard. The system employed various sensors such as temperature sensor, leaf wetness sensors, and humidity sensors in the vineyard. These sensors send the data sensed to the database in the ZigBee server which is linked to the sensors. Deployment of Wireless System Network (WSN) in any field needs to satisfy certain criteria and Zigbee alliance has developed open global standards called ZIGBEE. Zigbee compliances of four layers namely physical layer, medium access control layer, network layer, application layer as stated. The three devices: Zigbee Co-coordinator (ZC), Zigbee Router (ZR), Zigbee End Device (ZED) has different function in the WSN. Kalaivani *et al.* (2011) discusses end to end approach of Zigbee in agriculture. The server will store the data. The server is commissioned with a hidden Markov model algorithm in it. This algorithm is present to train the normal data sensed by the sensors and report any aberration in temperature, humidity or leaf wetness which can result in grape disease to farmer via SMS. Machine learning is blended in the system beforehand for astute deduction of disease in grapes. The additional advantage of this system is it also suggests the farmer pesticides and pacifies manual effort in the detection of disease.

3. IIS (Intelligent Irrigation System)

In one of the research conducted in Ankara, Turkey, implementing IIS (intelligent irrigation system), numerous positive perks were observed such as less moisture and temperature stress on soil, efficient water consumption, and neglecting human intervention in case of flood irrigation. The developed system works on three units such as Base unit (BU), Valve unit (VU) and Sensor unit (SU). The whole system is powered by solar panels. After the successful installation of every unit, BU will send the address to which the data is to be sent to SU. Sensors from the SU will sense the moisture content and send the detected data to a specific address in the BU. If required, BU will send a signal to VU so that it can calibrate the position of the valve in order to provide the soil with water. However, site-specific use of automatic irrigation system took birth in the early 21st century; this method proved to be a significant success as it reduced the cost, feasibility, and complexity of the developed system. Furthermore, the unit can be set up to transport the fertilizers and pesticides in the field using the same method. For that, new kind of sensors would have to calibrate for transmitting accurate information (Dursun and Ozden, 2011).

4. e-Agriculture Application

In 2016 a group of researchers came up with e-Agriculture Application based on the framework consisting of KM (Knowledge base and Monitoring) modules. The systems developed in IOT and Cloud Computing emphasizes on reliable architectures to provide timely information from the field over 3G or Wi-Fi. TI CC 3200 (RFID) launch pad was used to build the prototype with other necessary devices. Knowledge base has advantage over conventional IOT based systems; Knowledge Base is constructed to store vast complex structured and unstructured information to assist farmers or even an individual with no prior knowledge of farming. But finding right information in an appropriate manner is difficult where providing relevant knowledge should be distributed not only in an organised and complete manner, but also in absolute way. The knowledge based infrastructure allows adapting the changes in agriculture for a better extension and adding advisory services (Mohanraj *et al.,* 2016).

5. CNN (Convolutional Neutral Network) in Weed Management

As degree of automation is required in each and every field so the human intervention becomes less and it is very important to design a layout in the early stages of the mechanics and electronics. Weed management is the issue which farmers face a lot and computer vision can help to solve the issue. There is particular difference between a weed and the desired crop. CNN can help to distinguish among them and notify us to cut only the unnecessary plant. CNN has many algorithms which can even be used to identify plants and get the data accordingly for plantation.

6. R-CNN (Region Based Convolutional Neutral Network) for Fruit Detection and Counting

R-CNN extensively used in object detection and in automation it is used for fruit detection and counting of fruits. They have used VGG16 NETwith 13 convolutional network and also ZF network which has 5 convolutional layers. Data augmentation is used because it helps in artificially enlarging the dataset and changing the variability of the training data. The results discussed by them are promising as for both mangoes and apples. Faster R-CNN outperformed the ZF network approach. Cloud based decisions and support in the agriculture is booming now a days.

7. The Decision Support and Automation System (DSAS)

The Decision support and Automation system (DSAS) helps the farmers to control all the applications through its web portal. DSAS as different stages where to interconnect many devices on the single time and give the real time data to the farmer. The farmer plays the vital role as he can monitor the real time data and also control all the machines through software's. Systems like spray controller will spray the pesticide on the field in a defined amount. Similarly, irrigation controller helps to manage irrigation and fertilizer controller manages fertilizer. DSAS works through the data given by different sensors like soil moisture sensor, nitrogen sensor, *etc.* (Tan, 2016).

8. Wireless Sensor Based Irrigation System

Kumar (2014) used fertility and pH meter to take out the percentage of ingredients of the soil and developed wireless sensor based drip irrigation system. Ingale and Kasat (2012) used IC 89c52 microcontroller to build a smart irrigation system. The prototype supplies water only when humidity and moisture drops below a standard decided value hence it conserves water to a certain extent. A semi-automatic irrigation system was developed and tested on the field of Okra crops (*Abelmoschus esculentus*). The system used four moisture sensors and PIC16F877A processor was used. The valves in the system turns ON only when there is a voltage drop across any two sensors in the filed drops below a fixed value and remains ON until the value comes to the decided threshold value.The system designed by Encinas *et al.* (2017) monitors the quality of water by deploying state-of-the-art automation techniques. Shekhar *et al.* (2017) developed a technology which allows machines to communicate themselves.

9. Lentils Machine Vision

A Flatbed scanner was used as a hardware component with a pentium CPU. Various methods including K-NN, neural networks *etc.* were used for colour grading of lentil crops. Additionally an online neural classifier was used which gave almost 90 per cent accuracy in grading of crops. Cost effective and smaller scanner can be used instead of a large scanner which can give only required information required for lentil classification. Owing to the size of scanner, it gave variety of information which is less useful for the purpose of classification and grading. Canada (Shahin and Symons, 2001) industry has received tremendous advancement. Using deep learning is an added advantage over machine learning and it adds depth to machine learning. Many significant methods can ensure the farmers with better crops and proper field management. This in the end helps in overall growth of the country as food is the foremost need of any human being. IOT marked its significance to help in the real-time monitoring of the data. IOT is mainly used in an intelligent watering system. Effective use of the available fresh water is essential and with the advancement in the technology and application of automation water crisis can be solved. Traditional methods in agriculture have minor effects in this modern world. Water scarcity and flooding both are the major problems farmers are facing using the traditional approach. Many loop holes in this systemand the alarming need to protect the agricultural land leads to the development of agriculture automation.

Chapter 13

Technologies Used in Automation for Farmers

The primary goal of farm automation technology is to cover easier, mundane tasks. Here are some major technologies that are most commonly being utilized by farms.

1. Harvest Automation

Harvesting fruits and vegetables have always proven to be a difficult problem to automate. Harvest robots must be gentle with the produce to avoid bruising and damage. Agrobot has successfully developed the first robot for gently harvesting

Agrobot.

strawberries, no matter where and how they are grown. From a flexible mobile platform, up to 24 robotics manipulators work together to pick the fruit by which meets the farmer's quality standard.

Another company, Abundant Robotics, is the world's first commercial robotic apple harvest. Their machines handle fragile fruits by using a vacuum instead of any claw or hand-like graspers to pull apples from the branch.

Abundant Robotics.

2. Autonomous Tractors

Autonomous tractors can be controlled remotely or even pre-programmed to give full autonomy to a producer. Rabbit Tractor's autonomous tractor delivers value to row crop farmers not just through a reduction in labor costs, but through increased efficiency across operations and increased yield. Tractor automation

Rabbit Tractor.

kits are even being developed by Bear Flag Robotics that makes automation more accessible for farmers by affordably retrofitting existing tractors with cutting edge driverless technology and implement control.

3. Seeding and Weeding

Robotics developed for seeding and weeding can target specific crop areas. In seeding, this can easily reduce labor and mundane tasks on the farm. Weeding robotics can be incredibly accurate and reduce pesticide usage by 90 per cent with computer vision. Blue River Technology employs computer vision and robotics technologies to precisely spray herbicides only where needed and with exactly what's needed. This gives farmers a new way to control and prevent herbicide-resistant weeds. EcoRobotics is another company that produces a weeding robot – this is the first ever completely autonomous machine for a more ecological and economical weeding of row crows, meadows, and intercropping cultures.

Blue River Technology.

4. Drones

Drones can be used to monitor conditions remotely and even apply fertilizers, pesticides, and other treatment from above. They can also quickly and cost-effectively identify problem areas with imagery and infrared analysis to help farmers diagnose issues early on. American Robotics is developing a fully autonomous "Robot-as-a-service" with an autonomous drone, base station, and analytics platform that provides insights to growers at resolutions, frequencies, and speeds never before possible.

Robot-as-a-Service.

5 Autonomous and Robotic Labour

Replacing human labor with automation is a growing trend across multiple industries, and agriculture is no exception. Most aspects of farming are exceptionally labor-intensive, with much of that labor comprised of repetitive and standardized tasks–an ideal niche for robotics and automation. We're already seeing agricultural robots–or AgBots–beginning to appear on farms and performing tasks ranging from planting and watering, to harvesting and sorting. Eventually, this new wave of smart equipment will make it possible to produce more and higher quality food with less manpower.

6 Weeding and Crop Maintenance - Bonirob Robot

Weeding and pest control are both critical aspects of plant maintenance and tasks that are perfect for autonomous robots. A few prototypes are already being developed, including Bonirob from Deepfield Robotics, and an automated cultivator that is part of the UC.

The Bonirob robot is about the size of a car and can navigate autonomously through a field of crops using video, LiDAR and satellite GPS. Its developers are using machine learning to teach the Bonirob to identify weeds before removing them. With advanced machine learning, or even artificial intelligence (AI) being integrated in the future, machines such as this could entirely replace the need for humans to manually weed or monitor crops. The UC Davis prototype operates a bit differently. Their cultivator is towed behind a tractor and is equipped with imaging systems that can identify a fluorescent dye that the seeds are coated with when planted, and which transfers to the young plants as they sprout and start to grow. The cultivator then cuts out the non-glowing weeds.

While these examples are robots designed for weeding, the same base machine can be equipped with sensors, cameras and sprayers to identify pests and application of insecticides.

These robots, and others like them, will not be operating in isolation on farms of the future. They will be connected to autonomous tractors and the IoT, enabling the whole operation to practically run itself.

7. Harvesting from Field, Tree and Vine

Harvesting depends on knowing when the crops are ready, working around the weather and completing the harvest in the limited window of time available. There are a wide variety of machines currently in use for crop harvesting, many of which would be suitable for automation in the future. Traditional combine, forage, and speciality harvesters could immediately benefit from autonomous tractor technology to traverse the fields. Add in more sophisticated tech with sensors and IoT connectivity, and the machines could automatically begin the harvest as soon as conditions are ideal, freeing the farmer for other tasks.

Developing technology capable of delicate harvest work, like picking fruit from trees or vegetables such as tomatoes, is where high-tech farms will really shine. Engineers are working to create the right robotic components for these sophisticated tasks, such as Panasonic's tomato-picking robot which incorporates sophisticated cameras and algorithms to identify a tomato's color, shape and location to determine its ripeness. This robot picks tomatoes by the stem to avoid bruising, but other engineers are trying to design robotic end effectors that will be capable of gently griping fruit and vegetables tight enough to harvest, but not so hard that they cause damage. Another prototype for fruit picking is the vacuum-powered apple picking robot by Abundant Robotics, which uses computer vision to locate apples on the tree and determine if they are ready to harvest.

These are only a few of the dozens of up-and-coming robotic designs that will soon take over harvesting labor. Once again, with the backbone of a robust IoT system, these agbots could continuously patrol fields, check on plants with their sensors and harvest ripe crops as appropriate.

8 Drones for Imaging, Planting and More

What farmer wouldn't want a bird's eye view of their fields? Where once this required hiring a helicopter or small aircraft pilot to fly over a property taking aerial photographs, drones equipped with cameras can now produce the same images at a fraction of the cost. In addition, advances in imaging technologies mean that you're no longer limited to visible light and still photography. Camera systems are available spanning everything from standard photographic imaging, to infrared, ultraviolet and even hyperspectral imaging. Many of these cameras can also record video. Image resolution across all these imaging methods has increased, as well, and the value of "high" in "high resolution" continues to rise.

All these different imaging types enable farmers to collect more detailed data than ever before, enhancing their capabilities for monitoring crop health, assessing soil quality and planning locations to optimize resources and land use. Being able to regularly perform these field surveys improves planning for seed planting patterns, irrigation and location mapping in both 2D and 3D. With all this data, farmers can optimize every aspect of their land and crop management. But it isn't just cameras and imaging capabilities making a drone-assisted impact in the agricultural sphere–drones are also seeing use in planting and spraying.

9. Planting from the Air

Prototype drones are being built and tested for use in seeding and planting to replace the need for manual labor. For example, several companies and researchers are working on drones that can use compressed air to fire capsules containing seed pods with fertilizer and nutrients directly into the ground. DroneSeed and BioCarbon are two such companies, both of which are developing drones that can carry a module that fires tree seeds into the ground at optimal locations. While currently designed for reforestation projects, it's not hard to imagine that the modules could be reconfigured to suit various agricultural seeds. With IoT and software for autonomous operation, a fleet of drones could complete extremely precise planting into the ideal conditions for growth of each crop, increasing the changes for faster growth and a higher crop yield.

10. Crop Spraying

There are also drones currently available and in development for crop spraying applications, offering the chance to automate yet another labor-intensive task. Using a combination of GPS, laser measurement and ultrasonic positioning, crop-spraying drones can adapt to altitude and location easily, adjusting for variables such as wind speed, topography and geography. This enables the drones to perform crop spraying tasks more efficiently, and with greater accuracy and less waste.

For example, DJI (Da –Jiang Innovations) offers a drone called Agras MG-1 designed specifically for agricultural crop spraying, with a tank capacity of 2.6 gallons (10 liters) of liquid pesticide, herbicide or fertilizer, and a flight range of seven to ten acres per hour. Microwave radar enables this drone to maintain correct distance from the crops and ensure even coverage. According to DJI, it can operate automatically, semi-automatic or manual. Working in conjunction with other agbots, crops identified as being in need of special attention could receive a personalized visit from the nearest drone at the first sign of trouble. Being able to provide individualized attention to any part of the field as soon as it's needed could help to stop many problems before they spread.

11. Real-Time Monitoring and Analysis

One of the most useful tasks drones can take on is remote monitoring and analysis of fields and crops. Imagine the benefits of using a small fleet of drones instead of a team of workers spending hours on their feet or in a vehicle travelling back and forth across the field to visually check crop conditions. This is where the connected farm is essential, as all this data needs to be seen to be useful. Farmers can review the data, and only make personal trips out into the fields when there is a specific issue that needs their attention, rather than wasting time and effort by tending to healthy plants.

Given that drones for agricultural use are still early in their evolution, there are a few downsides. Ranges and flight times are not as robust as many farms would need–currently, even the longest running drones max out at around an hour of flight time before needing to return and recharge. The capital expenses are also still quite high, as much as $25,000 USD per drone for something likes Precision Hawk Lancaster. Less expensive models exist, but they may not come with the necessary imaging or spraying equipment.

BI (Business Intelligence) shared their predictions that IoT devices installed in agriculture will increase from 30 million in 2015 up to 75 million by 2020. Under this trend, connected farms are expected to generate as many as 4.1 million data points each day in 2050–up from a mere 190,000 in 2014. This mountain of data and other information generated by farming technology, and the connectivity enabling it to be shared, will be the backbone of the future smart farm. Farmers will be able to "see" all aspects of their operation–which plants are healthy or need attention, where a field needs water, what the harvesters are doing–and make informed decisions. And this discussion has only touched on the tip of the proverbial iceberg with the focus on vegetative crops; there is an equal groundswell of smart technology adoption for animal husbandry, and many more drones and robots for every aspect of farming. If every farm in the country becomes a smart farm, reaching that 70 per cent increase in food production is a certainty.

Chapter 14

Private Companies Technologies in Automation and Robotics for Agriculture

Agriculture is one of our most important industries. It provides food, feed and fuel necessary for our survival. With the global population expected to reach 9 billion by 2050, agricultural production must double to meet the demand. And because of limited arable land, productivity must increase 25 per cent to help meet that goal.

Consider these factoids:

1. Major US farming conglomerates are buying foreign land and beginning to farm there citing lower overall cost.
2. China is buying land in Africa and sending skilled workers to supervise those new farms.
3. Farmers and ranchers the world over are transitioning to precision agricultural methods, *i.e.*, subdividing their acreage into many sub-plots, in some cases, right down to the individual plant/tree/animal thereby enabling increased productivity and lower overall costs.
4. Unmanned aerial vehicles are being used to map, observe, sense and spray.
5. Unmanned (or at least autonomous) ground vehicles are providing more precise movements and thereby enabling precision practices.

6. The US Bureau of Labor Statistics reports that 2012 median pay for farm workers was $9.09.
7. The US Bureau of Labor Statistics reports that there were 749,400 agri workers in 2012, down 3 per cent (25,000) from 2011.
8. 74 per cent – approximate number of crop workers in the US who were born in Mexico or Central America of which more than half are likely to be undocumented (according to Fortune Magazine).
9. Cropdusters have the 3rd highest fatality rate among professions in the US. 90 per cent of crop spraying in Japan is done using small unmanned helicopters.
10. ResearchMoz, in a 1/29/2014 report, projects agricultural robot market size to grow from $817 million in 2013 to $16.3 billion by 2020.

Thus the agricultural industry is in transition. And that transition differs country by country, state by state, region by region as well as by type of farming practiced, from primitive to conventional and precision to experimental. A little bit of everything is going on everywhere but the general trend worldwide is toward precision agriculture supplemented by advanced technologies including robotics. Many factors are precipitating these changes in addition to global population growth and the cost and availability of labor: the diminishing availability and increasing cost of water, political and regulatory procedures and hold-ups; limited tillable acreage; better, cheaper and faster technological automation products; and climate change, to name just a few.

Modern farmers and ranchers are already high-tech. Digitally-controlled farm implements are regularly in use. There are partially and fully automatic devices for most aspects of agricultural functions from grafting to planting, from harvesting to sorting, packaging and boxing. Farmers use software systems and aerial survey maps and data to guide their field operations. They also use auto-steer systems included in many new tractors (or buy kits that do the same thing) which follow GPS and software guidance. Some farmers are already transitioning some of their operations to full autonomy. Thus forward-thinking farm owners today may be able to skip over slow, incremental improvements and jump directly to robotic and autonomous automation. But are the robots ready?

Companies Mentioned Sorted by Primary Function

Harvesting and Tractors

Tractors do two things: provide guidance to the devices they are towing, and pulling power. Current tractors are huge and if they break down, the entire operation comes to a halt. Autonomous machines don't need operators and can operate around the clock. Thus tight operational windows can be achieved for seeding and other time-sensitive activities.

1. Agrobot
2. Energid
3. Clearpath Robotics
4. Autonomous Solutions
5. Wageningen UR
6. Agritronics
7. Kinze Manufacturing
8. Amazone-Bosch
9. AGCO Fendt
10. Rowbot
11. Robotic Harvesting

Planting, Pruning, Potting, Grafting and Nursery Operations

1. Harvest Automation
2. ISO Group
3. Helper Robotech
4. Conic Systems
5. Wall-Ye

Thinning and Weeding

1. Blue River Technologies
2. EcoRobotix
3. Vision Robotics
4. F Poulsen Engineering

UAS, Inspection, Data Collection and Data Manipulation

UAVs are only as good as the other precision ag equipment and systems; if there are no computers on the tractors or controllers on the implements towed, and if they can't talk to each other, UAS data collected is just pretty pictures.

1. Agribotix
2. PrecisionHawk
3. senseFly

Smart Implements

New implements incorporate advanced control systems and can respond to commands from the towing tractor or provide their own mobility and navigation.

1. Jaybridge Robotics
2. Autonomous Tractor

3. Agrobotics AutoProbe
4. Naio Technologies

Agriculture is big business in every country around the world. Thus this is a timely review of the progress to bring robotic automation to an already automated industry. Bottom line: lots of activity, much of which will be coming online in the next year or two, but little market penetration thus far.

1. CropX Company

CropX Company.

It installs soil sensors throughout the field to alert growers when soil conditions are outside the ordinary. A farmer might receive a notification by there is lower moisture levels in a certain part of the field given that information in real time, the farmer has an opportunity to correct the issue (in irrigation systems), to produce higher-quality and larger yields. Similar IoT sensor technology has applications in storage safety as well. OPI (Open Prepress Interface) Systems' sensors for silos and elevators, for instance, track conditions and send alerts when heat or moisture might damage grain, or when a fire is possible. It's not hard to see the value of small, connected devices throughout the farm for increasing efficiency and safety. It's no wonder growers have taken to IoT so strongly. According to Business Insider, IoT device installations in agricultural settings are expected to increase at a 20 per cent annual compound rate over the next few years. With more accurate and timely data, growers can spend less time out in the field assessing and diagnosing conditions, and more time working on solutions.

The data from these IoT sensors, combined with information about seed, fertilizer, pesticide, rainfall and other factors, presents a compelling use case for software to optimize and predict growing conditions. Running a farm is a complex operation, with dozens of factors affecting every decision. As data collected from farms grows each year, software algorithms can become more complex in their predictive analytics, helping farmers decide when, where and how much irrigation, fertilizer, pesticide or other applications their farm needs.

Software can also automate the frustrating task of resource tracking and management. For instance, some software and apps enable growers to record all field applications and then track these resources from the field to multiple storage locations and then to the elevator. This resource management data can be incredibly powerful for optimizing agronomics on the farm, and it creates marketing value for food and beverage companies to differentiate themselves. IoT and farm management software are currently available and growing in popularity and sophistication. The next big frontier in agricultural automation is machines and robots that can do the work for humans. While most of these applications aren't fully commercially available or are too expensive currently for widespread use, many companies are racing to develop autonomous machines that are cost effective and reliably complete farm tasks throughout the growing cycle.

2. The DOT Power Platform

There have been significant advancements in automating tasks on the farm by smaller innovative companies. DOT Technology Corp's autonomous, diesel-powered seeder follows predefined paths to sow an entire field without any human intervention. Rowbot is another company that has developed autonomous vehicles that can apply nitrogen between rows of corn and even sow cover crops under mature corn late in the season, before harvest. In the UK, under human supervision, fully autonomous machinery planted, maintained and harvested a hectare of barley as part of the Hands Free Hectare project without a single human stepping in the field. While there were challenges with working in wet conditions and sacrificing some yield, the proof of concept project shows that it's theoretically possible to grow and harvest many commodities without human intervention.

Automation doesn't have to stop with tractors, and it isn't limited to grains. Drones are gaining ground as autonomous vehicles that can provide information about the health of crops from above. They can quickly and cost effectively to identify problem zones via imagery and infrared analysis and help farmers diagnose issues early on. Drones can also do some of the work typically left to airplanes or helicopters, dusting fields or even blowing water off ripe cherry trees after a heavy rainfall when the fruits are likely to burst.

3. Yamaha Precision Agriculture

It is testing a small, remote-piloted helicopter in Napa Valley that can spray fungicide. Vineyards in Napa often have tight row spacing and sometimes grow on steep hillsides. Until now, the only way to apply fungicide was for a farm worker to carry a backpack sprayer down the rows. Yamaha's helicopter can apply the fungicide from above, simplifying, improving, and speeding the application process. Harvesting fruits and vegetables also proves to be a difficult problem to automate, but several companies are up to the challenge. Harvesting robots must be gentle with the produce to avoid bruising and with the plant to avoid damage. Abundant Robotics has developed an early version apple picking robot that is gentle and precise.

4. Energid Company

It has a similar solution for picking oranges. Neither of these solutions is ready for primetime and the complexity of a real orchard. Time will tell if they turn into a viable, cost effective solution to a difficult problem. Every year, automation technology gets more sophisticated. What was cutting edge just a few years ago (guidance, drive-by-wire, continuously variable transmissions, remote sensors) has now become relatively commonplace and cost effective. Technological development toward full automation will only continue to accelerate as sensing, weather performance, terrain responsiveness and proactive autonomous decision making become more sophisticated.

It's not outside the realm of possibility that the coming decades could see farms managed from afar, with less need for human manual labour and more emphasis on human intuition, management and decision making. However, it's still clear that the human element of managing a farm is critical for the foreseeable future. Automation will enable farmers to scale their operations and be more efficient, but with all the complexities of weather and growing, it still takes human instincts and decision-making to run a modern farm (Automation is changing modern farming – ealAgriculture https://www.realagriculture.com › 2018/08 › automatio.).

5. AvanijalAgri Automation and ConserWater Technologies

AvanijalAgri Automation engaged in farm water and nutrient management by using mobile app.

The solution is powered by IoT. The solution saves water, manpower, electricity and helps in more crop yield.

ConserWater Technologies, a company headquartered in California, USA, and Avanijal Agri Automation, a company in Bangalore, have announced a partnership to help bring AI and IoT to the fingertips of Indian farmers. ConserWater Technologies is the company behind ConserWater, an AI tool that farmers can use to grow more healthy crops with less resource in a cost-effective and scalable manner. ConserWater uses satellites, weather, topography and other real-time data to help farmers accurately determine the inputs they need for their farm, without any expensive soil sensors systems. Using ConserWater, farmers in India have the potential to save as much as 50 per cent on their irrigation water use, while still growing healthy plants. AvanijalAgri Automation brings the IoT-based solution to the farmers to manage irrigation, nutrient of their crop and save water, manpower, electricity and grow more. Farmer can manage their farms completely using the mobile app. Together, AvanijalAgri Automation and ConserWater Technologies are creating new controller systems that are fully automated. Now, these 'smart' controllers can use the ConserWater AI to calculate irrigation needs, and can automatically irrigate the required amount of water and turn off by themselves, without requiring any additional work from the farmer. This integrated system not only enables the farmer to be efficient in his irrigation, but also significantly saves him time and effort. In many parts of rural India, electricity is unreliable and may be present only for a few hours each night. These controllers can automatically turn on when electricity is available, and can water efficiently using AI before shutting off, removing the need for the farmer to wake up and operate his pumps on his own late in the night. "We are very excited to work together with AvanijalAgri Automation to help farmers in India use less water and be better prepared for an increase in the prevalence of droughts due to global warming", said Aadith Moorthy, the founder and CEO of ConserWater Technologies. "We are happy to use ConserWater's AI powered Satellite imaging along with ground based sensing at the farm and save water for farmers which is the need of the hour in India", said Vijayeendra HS, director of AvanijalAgri Automation. Hopefully, in the near future, Indian farmers will no longer flood their fields with water, but will start using AI and IoT with Avanijal and ConserWater, and by engaging in precision agriculture, they will get better crops yield while saving precious resources!

Additional Information (ConserWater Technologies)

Website: www.conserwater.com

Press Kit: http://www.bit.ly/conserwater2

Facebook: www.facebook.com/conserwater/

Twitter: www.twitter.com/ConserWater

Additional Information (AvanijalAgri Automation)

Website : www.avanijal.com

Facebook:https://www.facebook.com/Avanijal/

Twitter: https://www.twitter.com/avanijal/

LinkedIn: https://www.linkedin.com/company/avanijal-agri-automation-pvt-ltd/###

Media Contact: Subramaniam Satyamoorthy Email: satyamoorthys@conserwater.com

Channabasappa Kolar Email: cbkolar@avanijal.com

Chapter 15

Top 30 Interesting Farm Automation Technologies and their Companies

1. AR Systems

This company claims to have developed a fully automated indoor farming system that can be managed from anywhere in the world. Judging from the picture, farms can be the size of a typical shipping container.

2. CNH Industrial

CNH Industrial is a huge name in the agricultural sector, and its autonomous tractor concept is appropriately epic. Futuristic and impressive, the driverless vehicles are probably beyond the means of most small farms.

3. Harvest CROO

This startup has developed a strawberry-picking autonomous vehicle. If it can engage in strawberry-picking effectively enough, it could solve all the arguments in the US about immigration, since many legal migrants cross the border to work in the agriculture sector.

4. EcoRobotix

A relatively lightweight vehicle, this solar-powered weeding robot can be used for weeding and reduce the reliance on pesticides, according to startup EcoRobotix, which raised $11 million to grow its business.

5. DeepField Robotics

Established as a business unit by industrial giant Bosch, DeepField Robotics would seem to have a bright future, by the admission of the people in charge – there are just so many solutions that could be developed for the sector.

6. Bosch

With its tremendous resources, Bosch is not only establishing business units like DeepField Robotics it's also developing farm robots like Bonirob within the company. Additionally, it's partnering with other companies.

7. Bayer

A bIoTech company which is in the process of acquiring Monsanto – a company associated with genetically modified farm foods. Bayer is working with Bosch to develop sophisticated, IoT-based farming systems that may reduce the reliance on pesticides.

8. Plenty

Indoor farming systems developer Plenty has received huge amounts of attention and investment and may become the first company to set up a chain of urban indoor or vertical farms which can compete with the likes of Whole Foods Market and other greengrocers.

9. Harvest Automation

This company developed a mobile vehicle designed around the needs of agricultural companies, but then it sold the technology to NextShift, which has broadened its application to industries beyond agriculture.

10. Drones

A variety of drone companies have been customising their vehicles for agriculture, including the largest drone maker, DJI, which is what the link is to. But while there is potential for their application, drones still have limitations such as flight time.

11. ABB and Heliospectra

Innovations in lighting are probably going to be critical to the success of indoor farming of many types, and multimillion-dollar businesses like Helispectra are partnering with industrial giants like ABB to make their greenhouse operations more efficient.

12. Autonomous Solutions Inc

ASI develops crucial technologies such as that which drives CNH's autonomous tractor. It also adopts other technologies, such as mesh networking, which may influence other companies in the industry to do the same.

13. Husky

Clearpath Robotics provides this unmanned guided vehicle that, in this instance, has been customised to deliver "precision irrigation" which, as the term suggests, can save significant amounts of water for farms and large-scale gardens.

14. Bühler and Microsoft

Although this moves us a bit further downstream, closer to the consumer, food processing is also seeing significant amounts of innovation, including this one where Microsoft and Bühler are working to reduce the amount of food that is wasted.

15. Fraunhofer

Cucumbers are among the most popular salad vegetables, and picking them presents challenges because they are soft. But if anyone can solve the issues, it's the boffins at the Fraunhofer Institute.

16. Gamaya

Sugarcane is one of the most valuable crops globally, and cultivation probably covers millions of acres. This company provides cloud-based software to monitor and manage large sugarcane fields.

17. Volvo

Volvo demonstrated a self-steering truck for sugarcane harvesting in Brazil. The automaker seems keen on such technology since it has developed similar self-driving technology for the mining industry as well.

18. Smart Ag

Another sign that computers are entering the agriculture sector in a bigway; Smart Ag offers cloud connectivity and monitoring services for driverless tractors and can, as far as we understand, fully automate regular tractors.

19. ZF

Zahnradfabrik Friedrichshafen is a big name in the traditional automotive world. A few months ago, the company demonstrated its concept for a driverless tractor and other related technologies.

20. ABB

The robotic arm being used on the Singaporean farm in question is not being perched on eggs and used to hatch them, as the headline might suggest. It's more a story about material handling, which is where robots are increasingly getting involved.

21. Universal Logic

Primarily a software company, Universal Logic has developed a robotic solution which is proving popular with strawberry growers, with Agricultural Robotics ordering 100 of its "Neocortex" systems.

22. Terrasentia

Many new innovations and great business ideas are emerging from universities, and this one is from the University of Illinois. It's a relatively small mobile robot and its makers have set up a company for it called EarthSense.

23. John Deere

Arguably the most well-known name in agricultural machinery, John Deere has invested a very large amount of money in Blue River's lettuce farming robot technology.

24. Spread

Spread was probably the first company to demonstrate a viable urban farming business model. It was so long ago that we can't find the story we did about it, so this is just a link to the company's website.

25. Indigo

This is another company we have yet to feature on the website, but that's just two so far. Interestingly, this company's systems hold out the promise of "pesticide-free farming", which is obviously something everyone can support.

26. Marijuana

The demon weed was always going to become a big business once it was decriminalised and legalised, and so it has happened that quite advanced farming technologies – particularly in greenhouse management – have been developed in the sector.

27. Abundant Robotics

Picking apples may sound like an idyllic way to pass the time, but as a full-time job in the heat of the summer, not so much. Hence this company's apple-picking robot could become a must-have for orchard owners.

28. Sensors and Chips

Tiny computers such as the Raspberry Pi and Arduino are being used to create highly sophisticated hardware-and-software systems that are within the purchasing power of many farms. The link leads to just one example.

29. Connectivity

As well as powerful chips and sensors available at low-cost, networking and connectivity technology – such as LoRa technology – is also available at much more affordable prices. All of which is leading to farms becoming IoT fields.

30. Fanuc

Another example of a "downstream" robotic system, this story proved quite popular as it tells of Fanuc robotic arms being used to process and pack lettuce for a Spanish company, enabling it to reduce its reliance on finding scarce human labour (Abdul Montaqimt, 2018).

Chapter 16

Robots in Farming Operations

From high-tech greenhouses to cloud seeding, here's how agricultural robots are helping farmers fill labor shortages – and our supermarket shelves. "High-tech farming" is no oxymoron. A contemporary agricultural operation is more likely to resemble Silicon Valley than American Gothic, what with apps that control irrigation, GPS systems that steer tractors and RFID-chipped ear tags that monitor livestock. Robots pick apples, gather strawberries, harvest lettuce and strip away weeds. Drones gather aerial images that help farmers quickly assess crop health. And robotic greenhouses are sprouting up thousands of miles away from traditional farmland regions, growing vegetables in the backyards of high-consumption urban markets. It all comes at a time when growers face a costly, long-term labor shortage and – with the global population expected to rise from 7.7 billion to 9.7 billion in just over 30 years – food demand is poised to rise significantly.

Here's how robots can help mitigate some of those challenges

1. Crop-Harvesting Robots

On its face, crop harvesting seems ripe for automation. It's physically taxing and highly repetitive and the kind of labor that's often most effectively targeted in the robot revolution. (See factories, manufacturing, mining, logistics processing.) But that's not necessarily the case. Picking crops also requires manual dexterity and a delicate touch. Many fruits bruise easily in the heat, and leafy vegetables are easily torn. And most robots just aren't advanced enough to handle that level of precision. Remember, it wasn't all that long ago those roboticists finally got a cutting-edge bot to catch a ball – simple enough for humans, far less so for robots. But agtech companies in the private sector and robotics departments in academia continue their efforts to clear that hurdle.

Crop-Harvesting Robots.

2. Harvest CROO

Location: Plant City, Fla.

A single Harvest CROO (Computerized Robotic Optimized Obtainer) machine will be able to pick a plant in just 8 seconds and also cover eight acres in a single day. The robot will reportedly be able to replace a team of more than 30 human pickers.

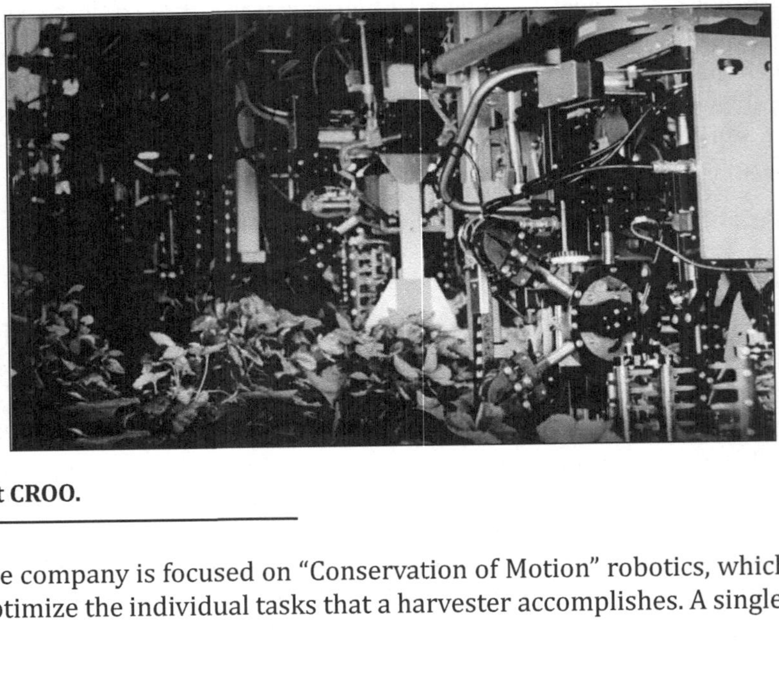

Harvest CROO.

The company is focused on "Conservation of Motion" robotics, which means they optimize the individual tasks that a harvester accomplishes. A single robotic arm isn't used to accomplish the picking of strawberries; instead the team decided to use multiple robotic components to separate the picking functions of leaf gathering, visual inspection, picking and then packing. Harvest CROO said that this approach allows them to build simpler robots, which run faster and operate at the scale needed to support current farm operations.

How it's using farming and agricultural robots: Gary Wishnatzki is vocal about the labor pinch he says growers have faced over the last few years. The owner of berry supplier Wish Farms told the *New Yorker* in April that he relies on workers hired through expensive temporary visa programs. Among reasons for America's labor shortage, the article identifies suppressed immigration and a diminished appetite among low-skilled domestic workers to do the backbreaking labor of strawberry harvesting. Wishnatzki's push to automate isn't about eliminating farm jobs, he argues, but meeting the demands of consumers who've come to expect fresh strawberries even in the dead of winter.

He's making that push as co-founder of Harvest CROO, a startup that developed an advanced strawberry-harvesting robot called Berry 5. It uses a variety of robotic components – rather than a single arm – to grab the leaf, pick the berry and pack it. Computer vision helps Berry 5 decipher ripe berries from non-ripe ones before plucking. And it's fast compared to human laborers, purportedly able to pick a plant in eight seconds and shift to the next in one-and-a-half.

Developed thanks to millions in investment dollars from others in the berry industry, Berry 5 is currently working Florida fields on a trial run. The company reportedly hopes to commercialize it before the end of 2019. Harvest CROO is a harvesting robot with sophisticated parts for grabbing the leaves, picking the strawberries, and packing them. It can distinguish between ripe and unripe fruits. The U.S. Bureau of Labor Statistics has recently reported an anticipated loss of six per cent of the agricultural workforce nationwide. In response to this trend, Harvest CROO Robotics has introduced a strawberry picking machine. In a 24-hour time span, the robot is reportedly capable of picking strawberries across eight acres of land.

With a reported 40 per cent of nationwide farm costs going to wages and other labor costs, companies engaged in efforts to implement AI into agriculture are on the rise. The Company claims that the machine's general workload capacity is comparable to thirty human farm workers. Based on the U.S. 2016 median annual wage for one agricultural worker, 30 workers would amount to approximately $676,200 in annual costs. Harvest CROO does not publish the cost of its robot on its website. However, if we consider the price tag of a competing strawberry picking robot at $50,000, it is plausible that the Harvest CROO robot could provide cost savings.

Harvest CROO Robotics.

3. Vegebot

Location: Cambridge, U.K.

How it's using farming and agricultural robots: Lettuce-harvesting has remained stubbornly robot-resistant thanks to the plant's fragile nature and close proximity to the ground. But researchers at Cambridge University made a breakthrough with their so-called "Vegebot," another computer vision-powered prototype.

Here's how it works: One camera scans the lettuce and gives a thumbs up or down for harvesting. A second camera (positioned near a blade) then guides the pick without crushing the plant. Meanwhile, a machine-learning algorithm "teaches" the robot to avoid unripe or diseased lettuce.

Vegebot hardly works with the speed or skill of human hands, but a series of test runs have purportedly established proof-of-concept success, which also augurs well for other above-ground fruits, vegetables and grains. "Every field is different, every lettuce is different. But if we can make a robotic harvester work with iceberg lettuce, we could also make it work with many other crops," Simon Birrell, of Cambridge's Department of Engineering, said in a press release. While a scalable commercial lettuce harvester remains elusive, efforts to produce one have ramped up. In 2017, for example, John Deere acquired CV agtech trailblazer Blue River Technology, which has made notable strides in lettuce-focused robotics.

4. Abundant Robotics

Location: Hayward, Calif.

California-based competitor Abundant Robotics is reportedly aiming to bring its commercial apple harvester to market by 2018. Co-founders Dan Steere and Curt Salisbury explain their vacuum-inspired technology. In May 2017, Venture Capitalist firm GV (also known as Google Ventures) was reportedly a lead investor in a $10 million Series A investment in Abundant Robotics. However, it is difficult to grasp an idea of the experience of Abundant Robotics clients. The company's website does not appear to provide client testimonials or case studies. Abundant is an advanced apple-harvesting robotic brand. It can identify the ripe ones, pluck them, and then store them. It works based on a vacuum-control mechanism and uses advanced algorithms.

Abundant Robots.

How it's using farming and agricultural robots: The apple-sucking vacuum robot from Abundant Robotics might initially appear brutish and imprecise compared to Cambridge's refined Vegebot, but the contraption's odd appearance – it resembles a tractor-affixed suction tube – belies some advanced tech. Employing sophisticated computer vision, the robot gulps up mature apples and bypasses their unripe brethren. It's also open to farmer assistance in that, as Wired notes, its algorithm can be updated based on feedback provided by ripeness-judging experts. Automated apple picking faces challenges similar to those of lettuce and strawberry picking. According to Abundant CEO Dan Steere, it "requires solving a number of complex technical problems in parallel, from visually identifying harvestable fruit and physically manipulating it to pick without bruising, to safely navigating the orchard itself." This past March the apple picker made its maiden commercial voyage, in the New Zealand field of fruit grower T and G Global.

5 Harvest Automation

Location: Groton, Mass.

Harvesting is not easy work for machines – they should be gentle enough not to damage fruits and vegetables. Nevertheless, harvest robots already exist and they cope with their tasks successfully. For example, Abundant is the first commercial company specializing in apple harvesting. The trick is to use a vacuum instead of graspers or similar devices. Agrobot, in its turn, is closer to the ground – the company builds robots to harvest strawberries. This innovation determines the fruit's ripeness via AI technology.

Harvest Automation.

How it's using farming and agricultural robots: All those charming decorative grasses, flowers and shrubbery accents at your local gardening center are big business. The commercial greenhouse market is poised to become a $38 billion industry within the next four years, and growers are increasingly using robots to help fill the high labor demand.

Founded by former employees of Roomba inventors' iRobot, Harvest Automation made its first product with this fast-growing market in mind. The behavior-based HV-100 robot handles the important, but highly repetitive and strenuous, work of spacing container crops and plants. (Greenhouse plants need space between them so they grow thick and bushy and resilient, but too much space means square footage isn't being optimized.) The HV-100 is built to keep running even in the scorching temperatures and less-than-pristine environments of nurseries that grow ornamental plants and specialty fruits and veggies.

6. Weeding Robots

If you've ever tended a personal garden, you're well aware that weed control is both important and difficult. Commercial agriculturists know it, too, but on a massive scale. Even when crop rotation is possible, many large outfits rely at least somewhat on the use of herbicides. But given the facts that plants can become resistant to weed killers and consumers are increasingly averse to chemically treated food, it's hardly a perfect solution. That's why weed-management robots – including ones that incorporate advanced AI to help distinguish between crops and weeds – are an attractive option.

Weeding Robots.

7. Naio Technologies

Location: Ramonville-Saint-Agne, France

How it's using farming and agricultural robots: One of the Old World's preeminent wine producers, Château Mouton-Rothschild is very New World when it comes to vineyard maintenance. The famed estate is among several vintners that have partnered with Naio Technologies to enlist Ted, the company's vine-tailored robotic weed killer. Electric and shaped like an inverted U, the long-running Ted simultaneously rolls over and around a vine row, using RTK satellite navigation to stay on course. (A drone maps out the initial plot of land that Ted surveys.) Industry-standard blades and finger weeders run along the base, pulling unwanted weeds from the vines and consequently decreasing the need for herbicides. Ted's robotic cousins include weeding robot Oz and vegetable robot Dino.

Naio Technologies.

8. Nexus Robotics

Location: Dartmouth, N.S.

Nexus Robotics.

How it's using agricultural robots: This scrappy Nova Scotia startup played David to some of North America's most prestigious robotics-program Goliaths at last year's Weed and Feed ag-bot competition. The company's top prize-winning entry, a weed-yanking autonomous robot dubbed R2Weed2 (yep, you read that correctly), employs artificial intelligence to differentiate between weeds and crops so the former are stripped and the latter left to grow. As it runs, R2 also gathers data that help farmers with soil analysis and environmental monitoring. A commercial version will reportedly be available later this year.

9. Robotic Greenhouses and Robot Farming

Instead of bringing robots to the field, one of the next great advances in farming automation will bring the field to robots. Below are two well-financed coastal startups –one West, one East – that are helping to plant the seed for a robotic-greenhouse future. Not everything is growable in this way, but for certain crops the improvements are striking. Both of these companies promise a dramatic decrease in the amount of water used (between 90 and 95 per cent less) for an equivalent crop yield, and both boast controlled indoor environments that eliminate the need for pesticides. Here's how they do it.

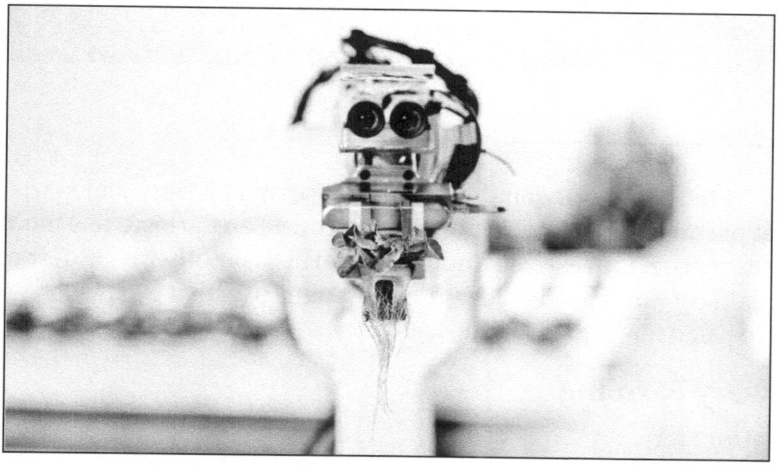

Robotic Greenhouses and Robot Farming.

10. Iron Ox

Location: San Carlos, Calif.

How it's using agricultural robots: Inside what the company describes as "the world's first autonomous farm," an 8,000-square-foot space that more closely resembles a research lab than a farm field, two cloud-connected robots oversee the growth of leafy greens (romaine and butterhead lettuce, bok choy, kale, arugula) and herbs (basil, cilantro, chives, sorrel, parsley), all grown inside heavy hydroponic pods. Using computer vision and sensors as its "eyes," one robot does the heavy lifting, transporting the pods across the facility; the second analyzes and picks the individual plants. It all happens beneath high-efficiency LED lights and under the watchful eyes of some dozen on-site robot and plant scientists. The Y Combinatorbacked company – founded by two Willow Garage alumni – claims the small space can yield as much as a one-acre traditional farm.

Iron Ox.

Iron Ox began production in its greenhouse-style facility and started selling at its first partner store, a market in its hometown San Carlos, a few months later. The long-term goal is to establish additional farms near other high-demand areas, which would cut out the long-slog transportation costs that exist when growing is isolated to just a few regions in the country.

11. Bowery Farming

Location: NYC

How it's using agricultural robots: Unlike the large, low-profile vats in which crops are nurtured at Iron Ox, Bowery Farming stacks up layers of trays, and each filled with greens, in the traditional vertical-farming format inside its Kearney, N.J., growing space. But it's also using robotics, artificial intelligence and LED to grow leafy greens and herbs with the same end-goal in mind: addressing the problems posed by labor scarcity, population booms and centralized farming. Here, a proprietary operating system and complex array of sensors collect data and maintain an ultra-precise balance of water, temperature, nutrients and humidity. At the same time, a team of vertical farmers help to harvest and watch over the crops.

The company – which has received investment capital from the likes of celebrity chef Tom Colicchio and Uber CEO Dara Khosrowshahi – sells its greens and herbs at Whole Foods, Foragers and Westside Market. In the New York metro area, they're also available through Peapod, Jet.com, Amazon and two of Colicchio's high-end restaurants.

Bowery Farming.

12. Aerial Imagery and Seed Planting Drones

Aerial imagery can save farmers a lot of time by giving them a bird's eye view of crops; that way, they can quickly get a sense of vegetation's health, insect issues, and irrigation layouts and weed growth. It even allows them to precisely determine how much pesticide the crops require.

Aerial Imagery and Seed Planting Drones.

Farmers can use a variety of subscription services to access these valuable flyover images (including thermal, infrared and NDVI) of their fields, but fewer companies have taken the full plunge into UAVs. That's most likely because of FAA restrictions on autonomous drones, which require that pilots be immediately ready to take control of a drone. Small unmanned aircrafts must also be kept within a would-be pilot's line of sight when airborne. But they're out there, if not yet in large numbers. Here are some companies that prove there's something's in the air when it comes to agricultural imaging, seed planting and cloud seeding.

13. American Robotics

Location: Marlborough, Mass.

How it's using agricultural robots: One of several Boston-area companies making notable agtech breakthroughs, American Robotics is the team behind Scout, an aerial imaging drone that fits the so-called "drone in a box" model. Between flights, Scout lives inside a weatherproofed box, where it self-charges and processes (via edge computing) all the data it collects. When it takes flight to examine fields, the box top opens and the fully autonomous drone lifts off, using artificial intelligence to plot and conduct the run. During missions which can be scheduled or launched on demand, Scout gathers crop stress data that farmers can use throughout a crop's life cycle. It's the kind of aerial surveillance that remains a key facet in the growing field of so-called precision agriculture.

American Robotics.

14. Taranis

Location: Tel Aviv

How it's using agricultural robots: Given the FAA's restrictions on autonomous drones, perhaps it's no surprise that some of the leading purveyors of agriculture-focused UAVs are headquartered outside the U.S. and caters to an international clientele. That includes Israel-based Taranis, which has brought its high-resolution scans to farmers in Europe, South America and North America. Along with more traditional aircraft, the company operates drones that use computer vision and data science to monitor crop stress and self-improve identification capabilities.

Taranis Drones.

15. UAV Systems International

Location: Las Vegas

How it's using agricultural robots: Any UAV pilot who wants to fly a drone that weighs more than 55 pounds at takeoff (including cargo) needs to petition for a special exemption through the FAA. That means non-exempt farmers who want to drone-scatter seeds over their acreage are limited in how many pounds of seed they can spread in a single run. Still, a number of manufacturers have developed drones marketed specifically for that purpose.

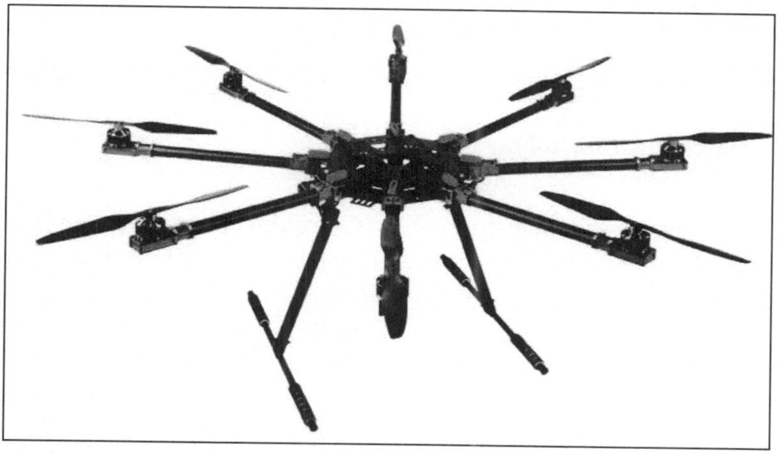

UAV Systems International Drone.

UAV Systems International sells two drones that spread seed and fertilizer, one with a payload capacity of about four pounds and another with a roughly 11-pound payload. Both have a two-mile flight range and a 20-minute limit, according to the company. UAV also markets crop sprayer drones and surveillance drones that inspect crop health.

16. Desert Research Institute Robots

Location: Reno, Nev.

How it's using agricultural robots: Cloud seeding is all about making it rain – literally. The concept dates back to at least 1946, when Dr. Bernard Vonnegut discovered that, under certain condition, introducing silver iodide particles to clouds could spur the creation of ice crystals, potentially precipitating precipitation. (Fun fact: Bernard was brother to literary legend Kurt Vonnegut.) But decades later, the process still has critics who note that variables abound and that it's tough to measure how much rain or snowfall is actually the product of a cloud seed. (The weather-modification practice also has military applications and, it should be noted, is internationally-banned for use in war). But as droughts worsen, government agencies keep trying. In July, Indonesia's disaster mitigation board used cloud seeding in an attempt to save drought-stricken crops.

Desert Research Institute Robot.

Today, clouds are seeded using airplanes, ground generators or gun- or rocket-propelled canisters, but scientists are also exploring whether drones can handle the task. In 2017, the Desert Research Institute – part of Nevada's higher education system – in conjunction with Drone America notably launched an unmanned cloud-seeding drone during an hourlong autonomous fight and beyond researchers' naked line of sight.

"Reaching this milestone," said Mike Richards, president and CEO of Drone America, "allows us to now focus on higher altitude, longer distance flights; as well as the extreme challenges of flying our advanced unmanned fixed-wing aircraft in the harsh, icy weather that come with optimal winter cloud-seeding conditions."

Chapter 17

Sensors and Automation: The Future of Farming in India

1. Cropping Sensors

Almost centuries Indian farmers follow a set of prescribed manual or guideless during fertilizer application and spraying. The local agri university or research agencies give farmers information on the dosage and method of application which is common to all farms and farmers. The evolution of crop sensors is going to change this scenario. Sensors tomorrow are going to provide site-specific information on the exact amount of nutrient required for a farm. This will save a lot of fertilizers and also conserve the precious soil resources.

2. Farming Vertically

Skyscrapers in agriculture are going to be the future in increasing farm output per area. With depleting soil fertility levels and increasing urbanization, land for farming tomorrow is going to be costlier than before. This means that food is going to be costlier and not affordable to common people. To avoid this situation, the vertically farming technique using soil-less medium and using artificial lights are the future. This technique will reduce cost, produce more, protect crops from bad weather and ultimately produce food which is safe and healthier.

3. Robotic Swarms

Just imagine hundreds of agri robots embedded with sensors collecting real-time information on farms. The data collected are shared to cloud and automatic reports are generated in terms of excess moisture, nutrient deficiency and many more insights of the farm. The robotic swarms are already on trial in Israel farms and are going to solve all untouched areas in farming. For instance, a global technical firm called FENDT is working on a project call MARS Mobile Agricultural Robot Swarms which is a Research in the field of agricultural robotics funded by the European Union.

4 Fitness Trackers for Cows

Fit bits have revolutionized human fitness; in the same path a start-up in India (StellApps) has introduced fit bits for cows. This technology used the IoT features in tracking a cow's sleeping time, eating and lameness period in helping farmers to take informed decisions.

5. Big Data in Farming

Collection of huge chunks of farming data, interpretation of the data and taking real time decisions will enable faster growth in agriculture. Data analytics and informed decision making will ensure greater levels of consistency, reliability, and accuracy. Information such as plant height and correlation with plant lifetime can provide insightful information in terms of an amount of care it needs to be taken to a particular farm.

Bringing the above technologies into middle-of-the-road adoption in Indian farms and making them affordable is the key to success. With the evolution of technology, research and rapid modernization in farming, the above technologies can be expected to be mainstream in Indian context by 2020 and financially viable for farmers by 2022. Artificial intelligence, self-driving cars, robotic vacuums... The last decade has given rise to phrases and concepts with the power to redefine how we live and work. While experts debate whether these new technologies are good or bad for society, the Institute of Electrical and Electronics Engineers (IEEE) suggests that automation and robotics have the potential to solve one of the world's most urgent and fundamental problems: food security.

While steam-powered tractors, cross-cultivation, and other agricultural advancements have minimized the risk of sharp famine for many nations that could quickly change. With the world population expecting to surpass 8.5 billion by 2030 and 9.7 billion by 2050, food demand will rise in parallel. Experts project that we will need 70 per cent more food than is consumed today by 2050.

We need an agricultural revolution if we hope to sustain our staggering numbers, and automation might just be the ticket. Farmers, researchers, and equipment companies have already begun experimenting with robotics in the field by developing machines that can boost agricultural production and efficiency across various stages of food production.

Here is a snapshot of some of the ways automation is already revolutionizing farming.

6. Robotic Pollinators

The robotics field is also making inroads in another key area of cultivation: pollination. The U.S. Department of Agriculture estimates that bees pollinate a full 80 per cent of flowering plants, including 75 per cent of the fruits, vegetables, and nuts that we consume each year. Unfortunately, bee populations have declined dramatically in the last decade.

Hand-pollination is possible, specifically with squash and a few other crops, but it is laborious and impractical. According to some researchers, fully autonomous robots could be the solution. These robots would be able to cross-pollinate plants with little-to-no human oversight. Some of these machines are flying drones that mimic insects while others, like West Virginia University's BrambleBee, are ground-based machines with robotic arms and tools that allow precision pollination. If deployed at scale, autonomous robotic pollinators could minimize the detrimental effects of colony collapse.

7. Farming with Drones and Intelligent Sprayers

While machinery helps farmers accomplish basic tasks faster, agriculture is still a very hands-on industry. Traditionally, if a producer wants to ensure a good yield, he or she must carefully monitor crops to identify and respond to signs of drought, disease, or poorly-fertilized soil. Experts hope to change that with strategically-deployed drones that can monitor crops on a wide scale and smart sprayers that can observe soil and plant health, water the crops, and apply chemicals or fertilizers as needed. These controlled environment agriculture (CEA) technologies allow speedy and targeted solutions to common agricultural woes, reducing herbicide and water use significantly.

Chapter 18

Success Stories of Startup in India on Agriculture Automation and Robotics

1. Nano Ganesh

Midnight in winter of 1981 in a small village in India – 81-year-old Sukhlal, a passionate farmer is sleeping with his family. Suddenly an inner alarm bell rings and Sukhlal wakes up and informs his wife, "I have to rush to the farm". His wife is worried about the dangerous path that Sukhlal has to take to reach the field at this time of night. Sukhlal takes his walking stick and a lantern and walks about 1 km through the dense bush. The fear of animals, snakes and other elements of the wild plays on his mind as he walks through the dark night. This is his daily routine. He does not know when he will return home.

The reason for Sukhlal's visit to the farm is to switch on his water pump. There had been no power for the whole day and when it was restored at 2.00 am, he had to walk to his farm to switch on the motor to irrigate his crop. Besides switching on the pump, he would have to wait for almost two hours to switch it off before returning home. The threats of theft of the water pump, cable or panel add to the difficulty in irrigating his farm.

a) Introduction

The story of Sukhlal is the story of most farmers in India who own water pumps. Convenient and safe pump operation, savings on labour and fuel expense and assuring family safety generated a need for a simple and low-cost solution to avoid frequent exhausting trips to distant water pumps. This involved controlling water pumps in a remote location and could only be satisfied by the revolutionary technology 'Nano Ganesh' which is an electronic hardware-based remote control solution for controlling water pumps with the help of a mobile phone.

Nano Ganesh is an electronic modem; it is a simple and low-cost solution derived from two decades of dedicated work in e-irrigation for empowering farmers, irrespective of gender, language, age or education.Twenty thousand farmers in India are benefiting from Nano Ganesh (as of August 2014). After installing the Nano Ganesh unit at the pump end, a farmer can switch it on or off with the help of a mobile phone from any distance. His phone also displays the availability of the power supply at the pump end as well as on/off status. Hence, farmers are not needed to physically visit sometimes hazardous pump sites in remote locations all the time.

Image Courtesy Farmer Controlling his Pump by Mobile Phone.

The basic version was developed in 2003-2004 and was continually upgraded over time. In 2009 the product was the Grand Winner at the 'Nokia Calling All Innovators' competition in the emerging market category in Barcelona, Spain. To understand the market scenario for rural automation in India, the scenario of the water pump industry has to be understood, because the demand for Nano Ganesh is a derived need from the demand for irrigation pump sets.

b) Market for Irrigation Pump Sets

About 25 million water pumps deliver water to Indian farmers. It is estimated that the total Indian pump industry is worth around US$980 million and this is expected to grow by around 7 per cent in the medium term. Domestic water pumps meet an estimated 95 per cent of the country's demand according to the FICCI report for the DST Lockheed Martin IIGP Program in 2011.

The Indian economy is primarily agrarian and the country is one of the highest exporters of food grains. Water pumps play a key role in agriculture and remote automation has made the irrigation process easier. The market for irrigation pump manufacturing is fragmented with many small and medium-scale enterprises (SMEs), a small number of large Indian players and a host of multinationals. The vast majority of smaller pump manufacturers have a local presence in states like Tamil Nadu, Gujarat, Punjab and Haryana, which are primarily agrarian economies. Their products find ready markets as purchase decisions are made by the farmers who are not familiar with emerging technologies. This is also one of the reasons for minimal product upgrade as far as agricultural pumps made by the SME sector is concerned. Things are gradually changing with improved marketing efforts, thanks to the entry of multinationals.

Over the past few years, domestic players have been looking at the huge global market. This has resulted in exports growing by around 10 per cent in recent years, and more importantly, a gradual improvement in quality, particularly amongst smaller players venturing into the global arena. Indian irrigation pumps are today exported to over 70 countries, including developing African nations. The Indian irrigation pump industry is expected to grow to US$1.25 billion by 2015, fueled mainly by soaring domestic demand from water and power sectors, complemented by an increase in international exposure.

c) Markets for Irrigation Technologies in Farming Communities differ in India owing to

- ☆ Education – illiterate, literate, advanced;
- ☆ Geographical locations – coastal belts, hilly areas, plains, deserts, jungles *etc.*;
- ☆ Cropping patterns – horticulture, cereals, rice, sugar cane, tea, cotton *etc.*;
- ☆ Family structure – joint, individual, nuclear *etc.*;
- ☆ Different types of irrigation technology;
- ☆ Irrigated and non-irrigated (monsoon-based) systems;
- ☆ Economies of scale – rich, medium and poor farmers;
- ☆ Social status – elites, marginalized, indigenous groups *etc.*;
- ☆ Available telecommunication connectivity;
- ☆ Population status – progressive, underdeveloped, dalit *etc.*; and
- ☆ Electricity and water sources.

d) Challenges Faced by Farmers in Controlling Water Pumps

- ☆ Distant locations of the water pumps in hazardous areas near rivers, wells, ponds, lakes *etc.*;
- ☆ Difficult terrain through dense plantations or cultivated areas;

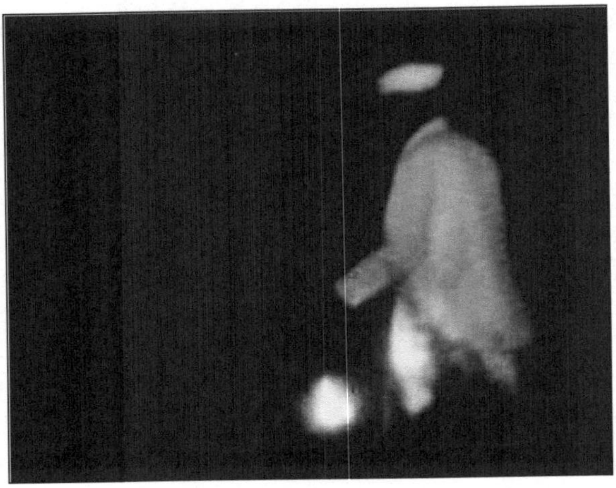

Image Courtesy Night Visits to a Water Pump.

- Dangerous wild animals on the way to water pumps
- Erratic power due to heavy density of electricity distribution in the irrigation zones;
- Odd hours or night operation schedules due to availability of power only during night hours;
- Electric shock hazards due to wet areas and electricity leakages near the water pump;
- Weather – rain, scorching heat, cold winters;
- Expenses for the operator such as fuel;
- Fear of water pump theft and accessories like cables, panels and starters *etc.* – replacement costs; and
- No quantification of usage and wastage of water and electricity.

Due to these challenges there is a tendency to avoid frequent trips to the pumps by continuously running them whenever electricity is available. They are not switched off after use. Different tools are adopted for running water pumps in continuous ON mode by using auto switches, automatic starters and sometimes by 'bypassing' the overload relays of the starter. This happens in sugar-cane producing belts to a larger extent.

Although it is understood that this is harmful to their water pumps and there is serious wastage of energy, the urgent need for water makes the risk of direct running of the water pump worth taking. But, due to the associated challenges, they are not switched off carefully and remain in the ON mode. There are about 25 million water pumps in India in the agriculture sector. Hence, there are huge losses of water and electricity as well as degradation of soil. Because of unavoidable visits to the pumps, the expenses for motorcycles, fuel and labour are a major burden for farmers.

e) Trends in Irrigation Automation

Two decades ago, many electronic devices like automated switches, timers, water-level controllers and indicators flooded the markets. They were simple and low-cost solutions to control the water pumps based on electricity availability, water level, dry running of pumps, seasonal need and so forth. Many local manufacturers dominated the market and very few branded products were available.

During the last decade, there was a major innovation in the irrigation sector, *i.e.* employing mobile phones for controlling water pumps. Since 2003, a farmer has been able to switch on/off water pumps with this technology. With some inertia in the initial period, the market accepted the technology after realizing the opportunity it provided. This concept was appreciated by the international community. There are many experiments that address such technologies for agriculture and it is perceived that tools should be low cost with the assurance of support at the local level. The need for an appropriate solution for remotely controlling water pumps arose from the challenges mentioned earlier. The result is the outcome of dedicated efforts and various trends in automation to solve the routine problems that farmers face during irrigation periods.

f) Challenges Faced in Implementing ICT Tools in the Rural and Irrigation Sector

- ☆ **Irregular electricity supply in villages and farms:** It is difficult to demonstrate, install, test and commission the products during the day due to erratic power supply.
- ☆ **Difficult terrain:** Field engineers sometimes face problems reaching pump sites for ICT installation owing to the rugged landscape which is difficult to navigate by motor vehicles at odd hours. Often they have to travel 5 to 10 km on foot.
- ☆ **Resistance of technicians to work in the rural sector:** There is a need for dedicated and passionate people to implement such technologies.
- ☆ **Serious hazards at the water pump sites:** Sometimes the pump area is wet which means electric shock threats. Also the presence of dangerous wild animals can be a deterrent.
- ☆ **Illiterate farmers and stubborn conventional mindsets:** Convincing farmers who are accustomed to their habitual irrigation routines to change to modern technology and inspiring confidence in it needs considerable patience by the promoter.
- ☆ **Heavy expenditure on advocacy:** Marketing, advertising, promotion, training, after-sale support *etc.* These expenses are a major drain on revenue from ICT products.

- **Investment for expansion:** Generating a sustainable business model to accommodate different expenses is a real challenge to convince investors. Currently a business model asks for investments which include either grants or long-term loans with low interest or demand advocacy by NGOs.
- **Low-cost needs:** There is huge disparity in the heavy cost incurred by R and D as well as promotion campaigns, the practical cost accepted by the farmers and break-even cost for entrepreneurs.
- **Buying inertia is high:** Time taken to percolate the product into the market. This does not marry with investors' financial calculations.
- **Business revenue declines in the summer and in the rainy season:** It is a significant challenge to arrange for specific business models to accommodate labour for optimum production in these periods.
- **Expenses in advocacy for energy and natural resources:** Water and energy are provided at subsidized cost or almost free; the value of these vital resources is not being acknowledged and there is a huge wastage. Major investment in terms of time and advocacy is incurred before promotion of products.
- **Competition due to local players in rural areas after the technology is accepted:** Although much time is invested in pioneering the technology, local competition rivalry starts when business revenues are accruing. This creates a considerable setback for the pioneers who have devoted their lives to building a strong market platform.
- Unpredictable purchase mindset of the farmers owing to unpredictable natural disasters, market rates, policies and so forth.

g) Mitigating Challenges and Establishing Optimism for ICT in the Irrigation Sector

In spite of these challenges, Nano Ganesh addresses them by reaching thousands of customers across the whole of India. Rural market characteristics are exponential, *i.e.* initially slow demand but great demand after trust is built. Technology should be low cost at least in the initial period to build trust among clients with assurance of support at the local level.

Despite the challenges there are success factors for ICT implementation in farm irrigation that should inspire rural and social entrepreneurs:

- Realization of the positive socio-economic change in their lives due to the technology acceptance is built through trust.
- **Practicality of the remote control system:** Once farmers experience drastic reduction in visits to pumps, saving money in terms of petrol and labour expenses, they will be more positive about the technology.
- **Local accessibility and after-sales support:** This is the most essential factor for maintaining a sustainable and scalable business.

- ☆ Demonstrations and pilot installations can help to 'win' farmers over to the technology.
- ☆ Low cost.
- ☆ Prompt service.
- ☆ A trusted brand.
- ☆ Commitment to provide three to five years service for the technology.

h) Description of Nano Ganesh

Nano Ganesh is a mobile phone-based remote control system by which a farmer can remotely control and monitor his water pump with the help of a mobile phone. Nano Ganesh serves as an interface between high voltage electrical starters and low voltage mobile phones enabling a farmer to switch on/off his water pump and check the availability of water supply near the pump end. Figure 6 shows the Nano Ganesh flow chart.

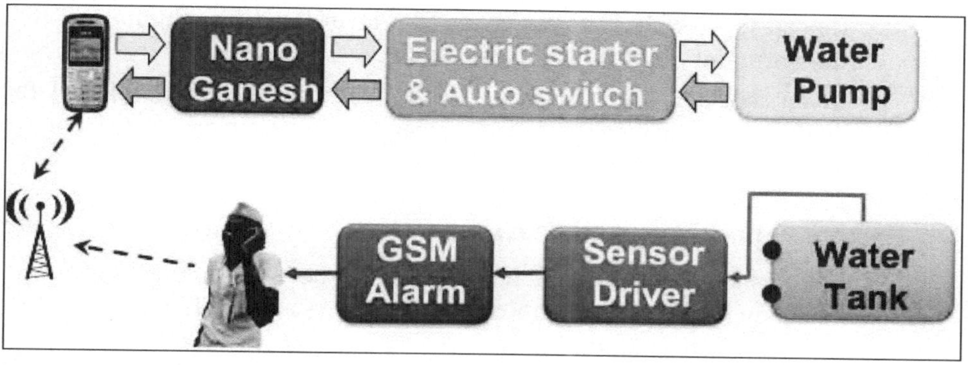

Figure 6. Nano Ganesh Flow Chart.

A farmer can operate the Nano Ganesh modem through an inbuilt mechanism for reception and feedback. By entering a preset code into his mobile, a farmer can switch on/off the pump from any location. Figure 7 shows the Nano Ganesh operation sequence.

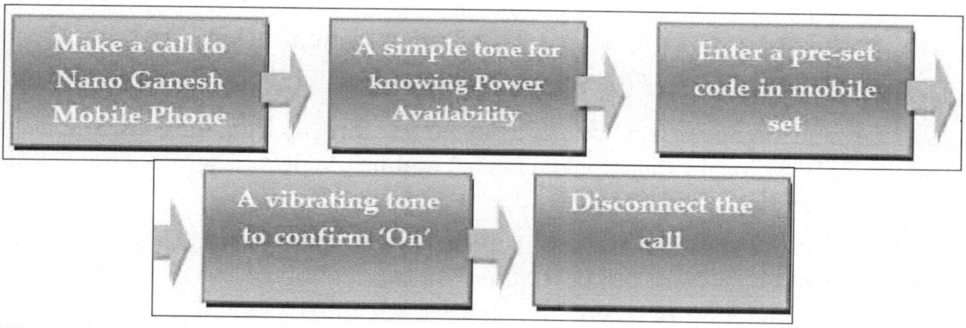

Figure 7. Nano Ganesh Operation Sequence.

In order to switch on the system, a farmer is required to make a call to the Nano Ganesh-connected mobile phone near the electric starter. The call will be automatically received. A long beep indicates the availability of power supply at the pump end. In case of no electricity supply, the farmer will not hear a different sound on his phone. Upon receiving the power-available indication, the farmer is required to press a preset digit code to switch the pump on. After dialing the code, a typical feedback tone confirms the successful running of the pump. After confirmation of successful running of the water pump, the farmer can hang up. To switch off the water pump, a similar process is adopted using a different preset code.

i) Functions of Nano Ganesh

- ☆ A farmer can switch his pump on/off with a mobile or landline phone from anywhere.
- ☆ He can check availability of the power supply near the pump end.
- ☆ He can acknowledge the on/off status of the water pump.
- ☆ He can receive alert messages about 'Power ON/OFF', 'Pump ON/OFF', theft attempts *etc.* on his mobile phone.
- ☆ In a recent developments water pumps are using Cloud technology for analytics of daily, weekly and monthly reports of pump operations, energy consumption, region-wise distribution of energy *etc.*

j) Unique Capabilities and Features

- ☆ Nano Ganesh is a pioneer technology that has been continuously upgraded based on feedback from thousands of farmers all over India since 1996.
- ☆ It is very simple and easy to use; it can be installed, repaired by local technicians.
- ☆ There is a range of products with different costs and applications (as low as US$12).
- ☆ No language, age or gender barrier for the user. An illiterate person can handle the system with ease. Housewives are effectively using this device to relieve their husbands from the responsibility of irrigation control.
- ☆ Nano Ganesh is an excellent tool in M2M applications where any electrical device can be connected to the telecommunication link with Cloud technology for computation requirements.

A farmer can purchase a Nano Ganesh modem directly from the company, local dealers, NGOs or local village technicians. A mobile phone and service connection has to be additionally purchased by the farmer.

Nano Ganesh has been exclusively designed for agricultural water pumps. Its product price ranges from US$12 for the basic model NGA to US$120 for the high-tech NG GSM model. More recent models are using the GSM module-based system with additional features to connect water pumps to Cloud technology.

Nano Ganesh - A	Nano Ganesh - AA	Nano Ganesh - ECO

$12	$17	$24

Nano Ganesh - B	Nano Ganesh - C	Nano Ganesh-GSM

$39	$50	$120

Nano Ganesh Model with a Mobile Phone.

Ossian Agro Automation is promoting different Nano Ganesh models designed according to the specific requirements of users and farming patterns in different site conditions, specific utility needs, budgets and various markets.

Nano Ganesh C is the most popular version. The product price is inclusive of input costs; indirect expenses on training, awareness programmes, advertising, travelling, promotions and professional fees.

k) Nano Ganesh C Model Product Details

Description	*US$*	*Per cent of MRP*
Product cost in US dollars	50	100
Raw materials	12	24
Labour	2	4
Testing	2	4
Logistics	16	32
Variable cost I/P per product (Sr.no.2+3+4)	7.5	15
Fixed expenses – training, advocacy, travels, promotions, advertising, salaries, others	23.5	47
Total product input cost (fixed +variable)	6.5	13

Description	*US$*	*Per cent of MRP*
Profit @13 per cent	30	60
Selling cost to state distributor	20	40
Discount in distribution – 3 to 4 levels	50	100

The installation cost is around US$5. This is borne by the farmer. In case of defects in the functioning of the modem, the Agro Electronic Commandos (AECs) charge a minimum referral fee of US$2. The AEC is a local village technician who is key to success in implementing electronics technology, especially in rural hazardous zones. He is a local man trusted by farmers as well as dealers have provides a useful bridge between the manufacturer, dealer and the end customer. Nano Ganesh management has always paid attention to ensuring adequate on-site and in-plant training for these technicians.

A Local Village Electrician is Vital for ICT Implementation in Irrigation.

l) Socio-economic Impact

Additional benefits to farmers and their families:

- ✰ Savings of sometimes US$500 per year;
- ✰ Flexibility in irrigation planning with precise control over irrigation. Every drop of water is used very effectively without visiting the water pumps;
- ✰ Enhanced quality of life due to better financial status;
- ✰ Saving of resources – electricity, water, fuel, time and labour;
- ✰ For a 5 horsepower water pump, Nano Ganesh saves 1 000 litres of water, 3-5 units of electricity, 2 hours of person time, 1 litre of fuel and 2 hours

of machine time per day. Soil quality is enhanced as vital minerals are retained – otherwise lost with excessive irrigation. Carbon footprint is mitigated;

- Farmers empowered in marketing and other developments;
- Enhanced water planning;
- Enhanced information dissemination;
- Heavy saving in the fuel and operator costs;
- Technology divide bridged; and
- Farmers and irrigation operators are made technology literate.

m) Specific Impact Related on Women

- Safe and secure with husband or otherwise at home all night long;
- Record keeping of daily irrigation;
- Updates on markets, money, weather *etc.*; and
- Employment opportunities –rural call centre, electronics assembly, rural marketing and training.

Precise Irrigation Control.

n) Economic and Business Opportunities

Additional income opportunities are created in the rural sector via Nano Ganesh such as installation, repair, courier services, training and demonstrations. A new business window has been generated for telecommunication companies. Nano Ganesh activity is changing the life of the Base of the Pyramid through different channels. Apart from the farmers, other members of society benefit either socially or economically from use of the technology via new dealers' networks, employment opportunities (assembly line wok for women), delivery and repair services and so forth

"Since we installed a Nano Ganesh mobile remote controller unit five years ago in our village, we can control the pump from any location: house, market, banks, farm, etc. Now our families can operate efficiently with a mobile phone. We save more than INR60 000 (US$1 000) every year on labour, water and petrol." Ravindra Bhujbal, Nano Ganesh owner from Sanaswadi, Shirur, Pune.

Mr Raghunath Patil Sonawane, Farmer and Water Supplier Pune, District Maharashtra State

Mr Sonawane came to our office ten years ago wanting a mobile remote control for his water pump very urgently, whatever the cost. Our engineers inquired about the reason for the urgency. We were shocked to learn that he had lost his brother in an accident on the highway en route to switch on his water pump. In fact, his brother had insisted about purchasing our product for some time. Site description: The source of water is near a river. The pipeline is installed along the highway and the water is collected in a water tank. But when the water tank is full, the source pump has to be visited by day or at night. It is hazardous during the rainy season near the river. Much water is wasted owing to negligence in switching the pump off when required. A motorcycle is used for such supervision.

To maintain the privacy of the users, all names have been changed. After remote control installation (2004): After installing our technology, Mr Sonawane stopped all trips to the pump. He could switch on/off his pump any time via his phone. He saved considerable amounts of water and electricity also on fuel and labour costs. Mr Sonawane observed, "This is a boon for farmers; now I depend on remote control for my irrigation activities. The system is easy to install and we do not have to rely on the company's technicians."

o) Village Organization

An Unnamed Grampanchayat, 17 Mawal, District, Pune, Maharashtra State

Site description: Grampanchayat water is pumped from a well to the overhead tank of the village. The approach road to the water pump is through the forest and hilly area and there is a fear of wild animals. Hence, the operators are scared to visit the water pump starters at night. Often electrical power supply is available only at night, so an operator needs to be accompanied by another person for safety reasons. The overhead tank must be filled completely before 05:00 for water pressure in the distribution pipelines to be adequate.

After Nano Ganesh installation (First Version 2003, Second version 2009): Mr Chavan, a reputed electrical contractor from Talegaon region installed our system with telephonic support from our company. Now, the same irrigation operator can control the water pump from his house at night too. Thus he is fresh enough for water distribution work in the morning. Another crucial benefit is that water overflow is addressed instantly. As soon as the overhead tank is full, an operator switches off the pump with his mobile phone straightaway. If the operator is away, he can control the pump from elsewhere because the remote control has virtually unlimited range. Mr Chavan notes, "All the grampanchayats must install this system, as drinking water is a precious asset which should be used very conservatively".

An NGO in Orissa, Orissa State

Site description: This NGO has resettled a tribal village from the forest near to a local village which is equipped with the basic facilities for standard living. However providing enough water for the additional numbers of people was a significant issue and water had to be lifted from a distant source in the forest. The approach road to the pump in the forest was too narrow. Nobody was prepared to visit this site at night due to fear of animals and the dark. Frequently electricity was available only at night. Hence, water scarcity was a serious issue as the water pump needed human operation.

After Nano Ganesh installation (12 units for different water supply schemes in 2012): Water distribution for different sites could be managed by a single person ensuring adequate supply of water at the proper time. There was no need to visit the sites as this could be accomplished remotely by the irrigation operator and money was saved on fuel expenses.

Mr Subhash Jadhav, a Farmer with an Artificial Right Leg, Beed District, Maharashtra State

Mr Jadhav was a formerly dynamic farmer whose water pump was about 1 km away from his field. But after an accident and loss of his leg, he was unable to carry out farming irrigation and had to depend on somebody else to control his water pumps. He also had to visit the hospital for leg therapy. His son was studying in the nearby town and was anxious about his father's condition. He went to the village to support his father but fretted about his career as well. After Nano Ganesh installation (2009): His son can control the pump from the town too. His father can also control the pump from any location (for instance from the hospital). In this case, Nano Ganesh also managed to help a family with personal issues.

Mr Prakash Patel, a Progressive Farmer with a Banana Plantation, Jalgaon District, Maharashtra State

Site description: Mr Patel once faced the problem of flooding on his farm for four to five days every fortnight. Despite ample water and electricity, sometimes he could not access the starter panel, as it was totally surrounded by water to a depth of 4 feet. So he used to run the pumps continuously with the help of automatic switches. Whenever power supply was available, the pumps functioned. But this led to huge wastage of water and electricity. Thus he had to use a small boat. Sometimes his assistant used to swim to the site to operate the starters. This case exemplifies the flooding problem that occurs in the backwater zones all over India in dam areas.

After installation of Nano Ganesh (2004): Mr Patel is now ecstatic as he has ample water and electricity and remote control over the pumps.

Nano Ganesh has proven that ICT tools can be effectively implemented in the agriculture sector. The pioneering work of Nano Ganesh has revealed the scope of various technologies and business models suitable for specific areas. The basic version of Nano Ganesh focused on solving the root problems of the farmer at the individual level. Advanced models can use Cloud and other technologies and hence the vast knowledge resources of the telecommunication and software sectors have been tapped for enhancing the overall system.

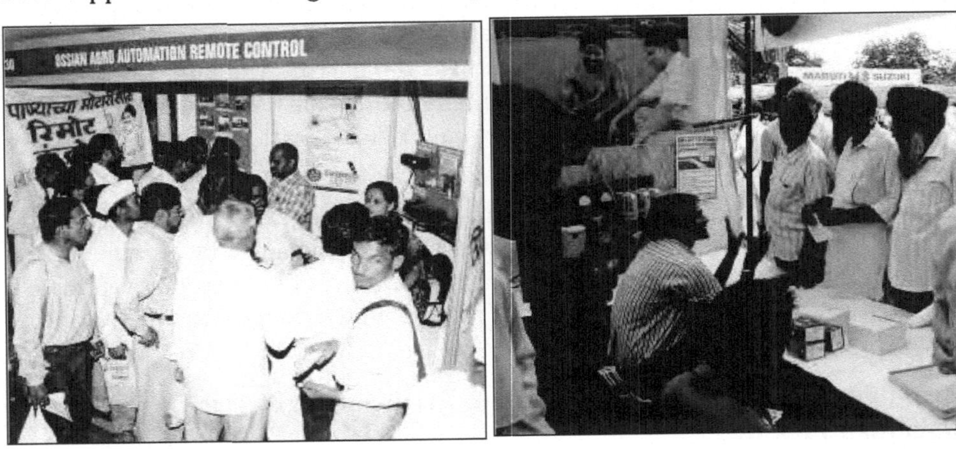

Advocacy of ICT at Rural Exhibitions.

p) There are Two Business Models for ICT in the Agriculture Sector

1. Traditional

Individual modems sold to individual customers through specific distribution networks. This business model addresses traditional farmers who buy ICT tools from the nearest available outlet, supported by trained technicians. The products are manufactured locally in-country by local businesses that can be backstopped by international institutions.

Cloud Technology for Water Pumps in the Rural Sector

Computation and analytics are possible with various telecommunication platforms. This business model can be driven by the government and international institutions on a larger scale for a country or a region. One state in India has declared Nano Ganesh as one of the major m-Governance™ initiatives as a strategy for addressing rural sector resources efficiently and effectively. The Nano Ganesh team has plans to serve as a catalyst for implementation of ICT for water pumps successfully on a global level. According to the Nano Ganesh team, expertise can be disseminated to local manufacturers for effective demonstration, installation and support services related to the technology.

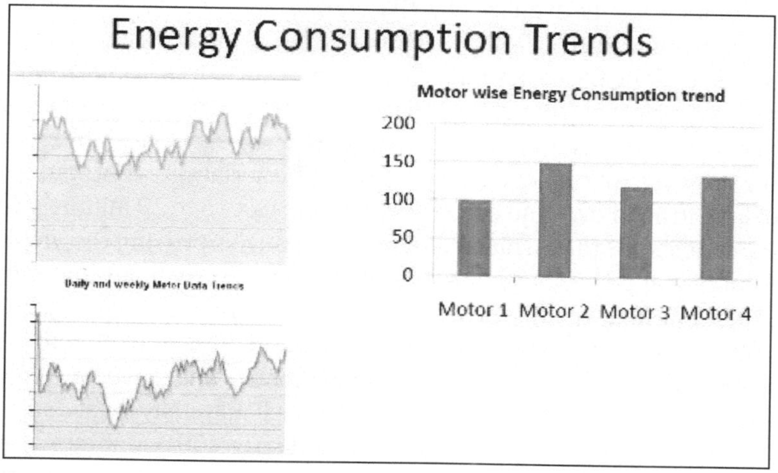

Analytics of Water Pumps Possible.

In countries like India, where agriculture is central to the national economy, adequate irrigation is crucial and the water pump is a core component. Nano Ganesh is committed to providing mobile connections for farmers and their families to enhance their quality of life. Despite the plethora of other developments in urban and industrial sectors, it is our duty is to empower farmers who bring food to our tables. It is the duty of international organizations to ensure that water, fuel and electricity in the irrigation sector (as well as elsewhere) are conserved to the fullest extent possible and wastage is avoided; this can be accomplished by helping farmers to adopt effective ICT tools like Nano Ganesh (Gerard Sylvester, 2015).

2. Thanos Technologies

Founded in	2016
Founders	Harish Alladi, Prathyush Akepati and Pradeep Palleli
Headquarters	Hyderabad, Telangana
Product Name	Agricultural Aerial Spraying
Technology Used	Aerial Robotics
Objective	To provide farmers with innovative aerial solutions to conventional terrestrial problems.

Startup Description

Thanos Technologies is an Aerial Solutions provider using drones to offer innovative solutions to conventional problems. Using a mix of Hardware and

Software platforms, the firm is redefining services which are conventionally catered to by terrestrial solutions. Thanos Technologies is a start-up that hopes to help farmers with its drones. The IIIT Hyderabad Centre for Innovation and Entrepreneurship and ICRISAT incubated start-up incorporates the finesse of aeronautics in making life easier for farmers. The team works primarily in the areas of agri-tech, agri-automation as an aerial spraying platform, agri-intelligence in terms of crop monitoring and plant count, apart from its more ambitious projects of working on army military and defense fields. The aerial spraying platform covers an acre of land in a uniform and efficient manner in just 15 to 20 minutes. This helps reduce input costs for the farmers, and also possibly improving the yield and soil health. A mobile app is used to mark the boundaries of the land that needs to be sprayed. Once marked, a couple of settings like the height and speed of the drone are entered. After this, all basic checks are performed and the drone lifts-off. No manual intervention is necessary during the spraying activity, and the drone comes back to the starting point by itself, or can be brought back manually if needed. Typically, the conventional method of spraying requires nearly 100-150 liters of pesticide (High-Volume Spraying) to be sprayed per acre. Apart from being incubated in CIE IIITH and ICRISAT, Thanos is also supported and mentored by RICH (Research and Innovation Circle of Hyderabad - A Telangana Govt. Initiative), and also by Surge Impact Foundation and Berkeley Smart AP. While quite a few companies seem to have either imported drones or built crude versions that are not market ready, thanos has it a point to design, build and test in-house a market-ready product. This is done to keep the capex low and the revenue model more financially viable. However, most of the components required building Unmanned Aerial System (UAS) platforms are not manufactured or even sold in India, and they have to be imported from China. Some of the materials we have imported include 3K carbon fibre sheets, DC motors, propellers and other minor electronic components. One of the critical components of the drone is the Flight Controller (the brain of the system), which is also being procured from China and other countries. The company

Thanos's In-house Built Spraying Drone.

is now working on high-endurance drone prototype (one hour or more flight time) that can be useful in surveillance applications for different types of users (Police, Army, Municipalities). They are also looking closely at drone delivery and AI-based solutions. The team has a spraying-as-a service model, where they intend to charge Rs 500 to Rs 600 per acre.

Active Regions	Telangana and Andhra Pradesh
Funding	Bootstrapped
Impact	25 farmers; area covered under commercial spraying: 180 acres
Website	www.thanos.in

3. AgriCx Lab

Founded in	2016
Founders	Ritesh Dhoot and Saurabh Kumar
Headquarters	Bengaluru, Karnataka
Product Name	Agricx Certification Software
Technology Used	AI and SaaS
Objective	To remove subjectivity out of the quality assessment process and make it reliable and easy-to-use, in order to enable a standardized and fair value (grades, price *etc.*) to different kinds of produces.

Startup Description

AgriCx provides a quick, accurate, portable and easy-to-use quality assessment tool. The solution currently involves imaging of produce using a standard smart phone and assessment using state-of-the-art deep learning and computer vision-based algorithms, making it extremely quick, portable and scalable. Agritech startup agricx is helping farmers get a fair price by grading and certifying their produce with its AI-based SaaS solution. Agricx, is striving to bring in a reliable crop-grading system in India, with the aim of eliminating middlemen in the supply chain and enabling farmers to earn a fair price for their produce. Ritesh Dhoot and Saurabh Kumar, who started Agricx in 2016, realised that the main reason for unfair pricing of agriculture produce was the fluctuating quality and standard of the produce, an acute problem, according to the duo. Agricx has developed an artificial intelligence (AI) and machine learning (ML)-based SaaS solution to simplify the grading system and eliminate variability in the quality of produce sourced. The core idea behind Agricx's business model is to streamline the procurement process by establishing accurate and precise grading systems with the help of technology. Agricx lab seeks to be a bridge between farmers and cold storages on one side and procurement specialists on the other, helping the latter procure their requirements in an efficient manner with the use of technology.

For this, the startup has partnered with cold storages (who are measured against stringent parameters) across India and listed their certified potatoes on its platform. Agricx also provides an additional offline service where their internal team visits cold storages to double check and evaluate the produce. However, this is an on-demand service and is provided over and above their technological solutions. The startup's core service is completely based on the SaaS application it has developed. The application enables cold storages to click images of samples and grading is done based on some pre-determined measures. The data is then stored for future reference.

Who graded the sample, the quantity of sample used for assessment, when and where the grading was done and the certification software uses an advanced proprietary algorithm to accurately detect grade specifications of potatoes from the images captured.

The application is also easily scalable as it can grade high volumes and different types of samples in minimal time, thereby establishing a reliable and transparent grading system. A single image of the produce can be evaluated just within 30 seconds. Agricx's USP is its deep learning technology. The startup's convolutional neural networks (CNNs/ConvNets)-based algorithm aid in image processing, classification, and segmentation of produce. For the uninitiated, CNNs are artificial intelligence networks that analyse visual imagery. In this case, they help determine the quality of produce by evaluating the crop images.

Mode of Action

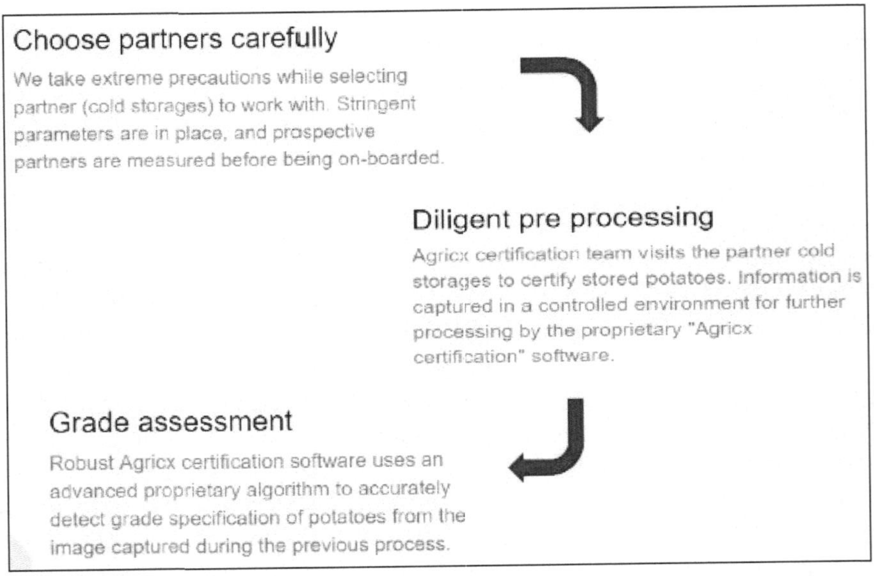

AgriCx Lab.

Active Regions	North India (UP, Delhi-NCR, Agra, Gujarat, Mohali *etc.*) and Karnataka
Funding	INR 3.25 Cr till now from the Centre for Innovation Incubation and Entrepreneurship (CIIE) and Ankur Capital
Impact	Clients such as Mccain India, Mahindra and over 100 cold storages in the country an also have already certified over 2 Mn Kgs of Potato in India
Website	www.agricx.com

AgriCx.

4. Fasal – Climate Smart Precision Agriculture Solution

Founded In	2018
Founders	Ananda Verma and Shailendra Tiwari
Headquarters	Bengaluru, Karnataka
Product Name	Fasal Microclimate System and SaaS
Technology Used	AI, Ml, IoT and Data Analytics
Objective	To increase and improve the quality of per acre yield

Startup Description

Fasal is a smart agriculture solution which when installed, starts recording variety of microclimatic parameters like temperature, humidity, air pressure, leaf wetness, soil moisture, soil temperature, wind speed, wind direction, and rainfall. It then uploads this data to our cloud server, where our predictive engine runs machine learning and AI algorithms to help farmers in:

Predicting crop diseases, and pest outbreaks, Ideal time to sow and harvest any crop, Optimizing resource utilization like water management, pesticides, fungicides, and fertilizers *etc.* For example, it informs farmers what the right time to irrigate or spray pesticides.

It provides real-time alerts based on farm, crop, and weather conditions to take preventive actions against reacting to situations by the farmer.

Online Farmer's network, where farmers all across India can ask questions, share new agriculture practices *etc.* learn about mandi rates as well as various government schemes.

Mode of Action

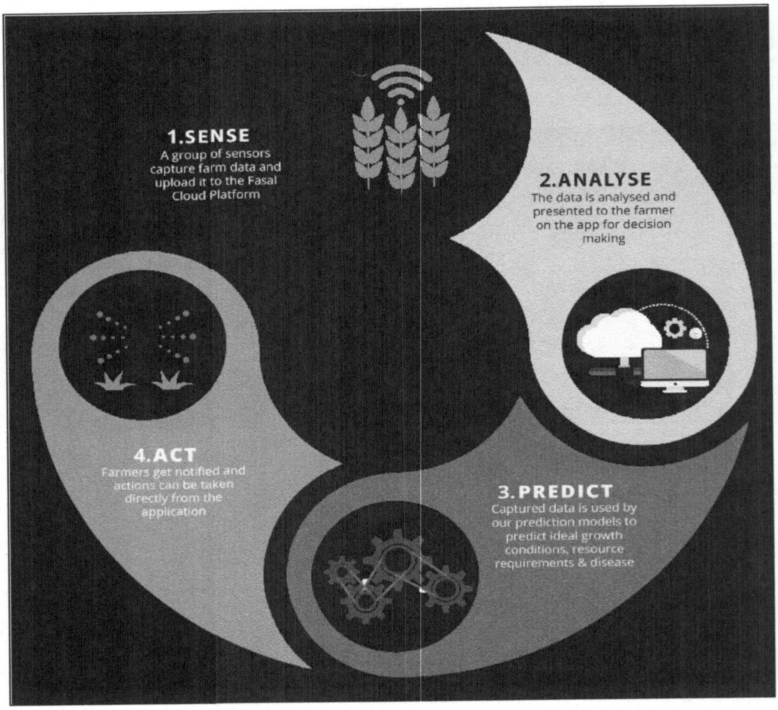

Fasal - Climate Smart Precision Agriculture Solution.

Step by Step How Fasal Works

1. **Sense:** Fasal monitors critical microclimatic parameters, including Temperature, Humidity, Rainfall, Soil moisture at various levels, Leaf wetness, Soil temperature *etc.*, 24x7x365 from your farm and uploads it to Fasal cloud platform.
2. **Analyse:** The data is then analysed and presented, making your crop's health accessible to you anytime, anywhere on any device for decision making like irrigation management, resource optimization, increasing yield, increasing quality of yield *etc.*
3. **Predict:** Captured data is used by our prediction models to predict the ideal growth conditions, resource requirements including fertilizers, crop diseases, and microclimatic weather predictions.

4. **Act:** Farmer gets notified on his device and actions can be taken directly from it, like switching on drip irrigation *etc.*

Their business model includes a nominal monthly subscription fee. There is no upfront charge or deposit. It is a pay-as-you-go model. For large scale and institutional farmers, they also offer to sell their IoT device and charge less subscription on software usage. It has worked pretty well for the startup to penetrate the Indian agriculture market which is very price sensitive. Fasal is competing against Indian startups like Yuktix and Exabit Systems, while the competition worldwide includes Cropx, Pycno, The Yield technology Solutions, and more.

Focus Crops	Tomato, Capsicum, Potato, Onion, Chilli, Pomegranate and Grapes
Active Regions	Bengaluru and Chhattisgarh
Funding	Friends and Family: $45,000, Zeroth.ai Pre-Seed Funding: $120,000
Impact	20 Independent Farmers, 4 B2B Farmers and Grover-Zampa, One of the Largest Vineyards of India is Fasal's customer
Website	www.fasal.com

Fasal – Climate Smart Precision Agriculture Solution.

5. FlyBird Farm Innovations

Founded In	2013
Founders	Satish KS
Headquarters	Bengaluru, Karnataka
Product Name	Siri – Digital Motor Starter, Siri Nano – 4Channel Controller, Siri – Sensor Based Irrigation Controller, Siri – Volume Based Irrigation Controller, Siri – Timer Based Irrigation Controller, Siri Link – Mobile App and Web Login Controller and Siri – Smart Nutigator - Fertikit

Technology Used	Micro-irrigation System, Web App/Mobile App Controllers, IoT, Timer Based Machine
Objective	To make every drop of water count with affordable, simple and reliable irrigation controllers for every farmer (marginal to large) which will save water and energy

6. vDrone

Founded In	2017
Founders	Kunal Sharma and Pranav Manpuria
Headquarters	Bengaluru, Karnataka
Product Name	Agricultural Aerial Mapping
Technology Used	Aerial Robotics and Mapping Technology
Objective	To provide farmers with actionable data through mapping technology

V Drone.

Startup Description

Started by two engineers, vDrone uses drones to take pictures of the field, which are then fed into the software developed by the startup and used to extract thermal images of the plants. Using innovative mapping technology, the Bengaluru-based startup quickly analyses areas of the farm which need attention, and helps the farmer cater to these needs. Thermal imaging is used to assess a crop's yield by determining the condition of the seeds, the soil, the cropping pattern, the weather, access to water, use of fertilisers, and the lack of pests. vDrone Agro develops custom-built drones to map farms for precision agriculture, and provides farmers with actionable data to improve crop health and increase yield.

The firm charges farmers Rs 400-800 per acre for drone deployment. It has developed a software which analyses the data collected to provide details to farmers about the health and quality of land and crop, such as which portions are less fertile, what portions have stunted growth, *etc.* The drones are lightweight multirotor quadcopters. The start-up has so far tested and deployed drones at farms in Karnataka, where over 500 farmers have used their services. It is now working with some agro-companies in Andhra Pradesh as well. Though the civilian use of drones has been banned by the directorate general of civil aviation, the founder said that vDrone Agro is one of six drone companies working with the regulatory body to set up a platform for such companies. With vDrone, farmers are able to get a better understanding of crop health by monitoring the data mapped through drones. The enterprise seeks to improve productivity and efficiency by permitting a more focused approach to farming through data. The International Centre for Entrepreneurship and Technology (iCreate), an incubation centre for technology businesses, helped vDrone with resources and mentorship. vDrone develops and designs drone-based systems for the agriculture sector with services like crop health data analysis, mapping, analytics solutions, *etc.* With the Rabi crop harvest just around the corner, these two engineers are now flying their drones across a five-acre field. Some might say what a drone can do on a farm other than take pictures of the field. But Pranav and Kunal have built software into their drone cameras which read thermal images of the plant making food. Once the drone lands, they collect the images from its SD card and start mapping the farm on their laptops. Next, they analyse the images and are able to tell the farm owner which areas of farm are fertile and which are not.

The startup works in 4 steps:

1. Order a Scan: Place an order through website, or via a call.
2. Deploy Drone: The firm plans the flight path and deploys the drone.
3. Scan Farm: The autonomous drone scans the land in minutes.
4. Actionable Data: The scan is processed to provide actionable data.

The startup has figured a business model that is based on the visualisation of the image and the recommendations on top of those images. Most of the recommendations will be about the energy produced in the plant and what should be done. The vDrone is built in Bengaluru and works on a pay load of less than 6kg. It is built on light materials and is a civilian drone.

Active Regions	Karnataka
Funding	Undisclosed
Impact	500 farmers have used the service in Karnataka
Website	www.vdrone.xyz

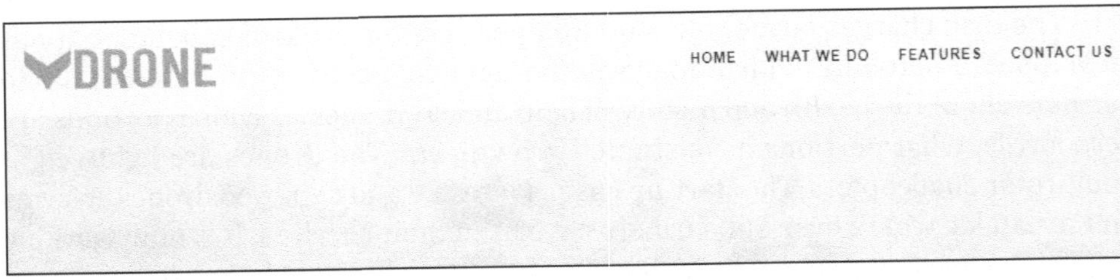

V Drone

7. Yuktix Technologies

Founded In	2013
Founders	Rajeev Jha
Headquarters	Bengaluru, Karnataka
Product Name	Automatic Weather Station, Air Quality Stations, Wireless Data Logger and Sensor Catalog
Technology Used	IoT and Wireless Sensing
Objective	Providing hardware to farmers to reduce the post-harvest losses by measuring indoor environment of Cold Storages and Warehouse

Startup Description

Yuktix Technologies was founded in October 2013. The firm has been working to create indigenous remote monitoring solutions for emergent markets. It provides data capture and communication solutions for remote monitoring and environment sensing. Yuktix weather stations collect meteorological information such as temperature, humidity, rain, wind speed every 15 seconds and up to 12 samples are taken for each transmitted reading. They provide data parameterized to conform to standards set by the Indian Meteorological Department (IMD) and the World Meteorological Organization. The station was designed with enough flexibility to be used as an air quality or water quality-measuring device simply by adding the pertinent sensors. The stations can serve as ready-to-deploy templates for smart cities and urban monitoring. The startup is currently dealing in three sectors, namely Agriculture, Urban and Industrial. In regard to the environmental aspect, agricultural indicators are used to assess the ways in which agriculture affects environmental quality and how changes in agricultural practices change the environment. 1) Crop Disease Prevention, 2) Smarter Greenhouse, 3) Post Harvest Monitoring and Remote Management can make any agricultural activity capable of producing more and keeping the environment safe for more production. Yuktix provides sensors based smart devices for climate control and precision agriculture which can measure climate variables inside the greenhouse and control light (Lux), humidity, temperature, water pumps, cooling pads and shade nets inside your greenhouse to maintain ideal climate conditions. Yuktix has developed a device along with a compatible sensor catalog to monitor environment variables inside greenhouses. The device can interface to Temperature, Humidity, Lux, Soil Temperature and Moisture, CO_2, Leaf wetness and soil pH *etc.* sensors. A greenhouse operator can monitor environmental conditions inside the greenhouse from anywhere on a mobile device. Yuktix also provides cellular and long range radio (wireless) communication options. The company is monetising by providing automated data. The team deploys the device to generate data and gives it to the user. Moreover, the services are charged for once during the rollout and then the company pays a recurring cost for using those services.

Mode of Action

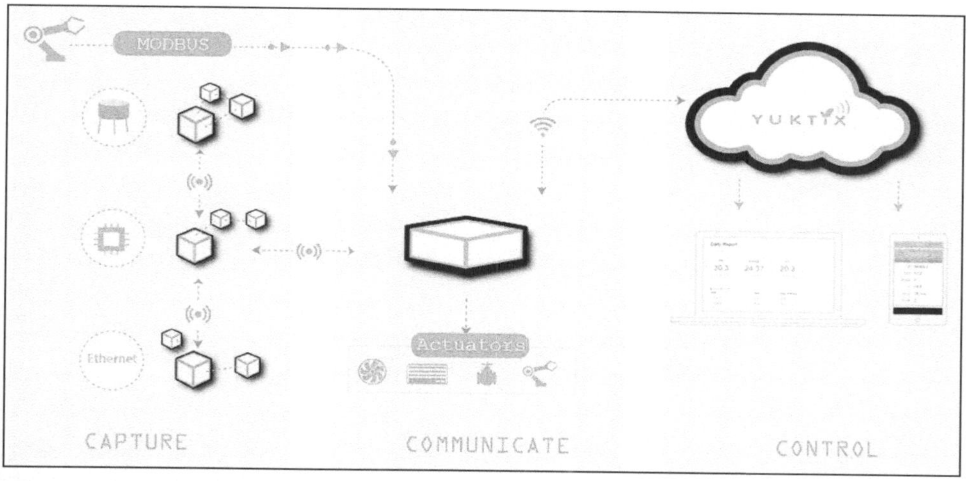

Yuktix Technologies.

Yuktix Technologies.

Active Regions	Bengaluru, Chennai, Hyderabad and Ooty
Funding	Undisclosed
Impact	-
Website	www.yuktix.com

Anupam Anand and Saravanan Raj (2019).

Chapter 19

Indian Agritech Startups to Watch Out for in 2018 on Automation and Robotics

India is moving at an accelerated pace towards digitization automation. The farm sector which employs about 58 per cent of India's workforce and contributes to about 14 per cent of the current GDP, has adopted technological advancement to warrant uniform progress. However, farm mechanization in India still has less than 50 per cent penetration. The Govt of India has been taking several measures and a significant drive towards this end is the multi-body project "Development of autonomous multipurpose agricultural robotic platform" funded by the Ministry of Electronics and Information Technology and run by C-DAC Kolkata as the nodal implementing institute along with IIT Kharagpur, BAU Ranchi and KCT Coimbatore. Under this project, IIT Kharagpur has designed a robotic system capable of identifying the plant diseases (say vegetables, maize etc.) through the camera-captured image analysis and spraying the appropriate pesticide, as the situation demands. The robotic system is a tracked mobile manipulator that aims to achieve the conflicting objectives of increased productivity, improved quality.

This robotic system consists of:

1. A tracked vehicle suitable for negotiating rough terrain field
2. A serial manipulator mounted on the tracked vehicle capable of holding the camera and pesticide spraying nozzle

3. A pesticide spraying module
4. Camera-based vision module

Prof. D K Pratihar, from the Dept. of Mechanical Engineering who is leading this innovative project said, "Our device can detect the plant diseases using camera and image analysis thus model the uncertainity associated with the human vision system and consequently remove it. This will help the farmers to detect the disease(s) correctly. Further use of such a device instead of manual operation of spraying pesticide will protect the farmers from related probable health hazards." The device has been fabricated by Sun Fab Industries Pvt. Limited, Mumbai (in collaboration with TECHNIDO).

"The serial manipulator will have sufficient dexterous workspace making it capable of spraying the pesticides from the top, bottom and various sides of the plant, which may be difficult to achieve utilizing the widely used drones," said Prof. Alok Kanti Deb from the Dept of Electrical Engineering. The developed robotic system is battery-driven one and once it is fully charged, it will be able to perform its duty in the field for about two hours. The robot will be operated remotely using buttons placed on a control panel and thus, a specialist technician may not be required for operating the robot in the field.

"We will soon resume the final assembly of the robotic system and field testing as we are recovering from the present COVID-situation. We have further plans in the future to make this robot autonomous and intelligent," confirmed Prof. Pratihar.

The Government of India is giving a major boost to innovative rural livelihood technologies. In May 2020, the Finance Ministry had announced an economic stimulus of 1.5 trillion to boost agricultural infrastructure and logistics. Talking about the potential of such technologies Director, Prof. V K Tewari, who himself is an expert in farm machinery design and precision agriculture said, "If we are looking at a comprehensive development towards Atmanirbhar Bharat, the strategic importance of such technology development is profound. Most importantly such technologies need to be taken from the lab to the land."

1. Aarav Unmanned Systems

An unmanned aerial vehicle startup incubated at IIT Kanpur with a team of designers, artists, developers, and engineers from IIT Kanpur, IIT Bombay, IISc Bangalore and IIST, Aarav Unmanned Systems aims to build the future of drones and their applications in the enterprise space. The startup's drones provide high-value engineering solutions to enterprises across GIS (geographic information system) surveying/mapping, industrial inspection and precision agriculture. Thus it holds the potential for making precision agriculture scalable in the country which is far behind in employing technology.

By providing accurate 3D representations of the terrain surface, it can transform and change the way decisions are made in agriculture through its solutions to optimise irrigation, fertilisation, pesticide distribution and early failure warnings. In February 2016, the agritech startup raised an undisclosed amount in Seed funding from StartupXseed Ventures, The Phoenix Fund, and other investors.

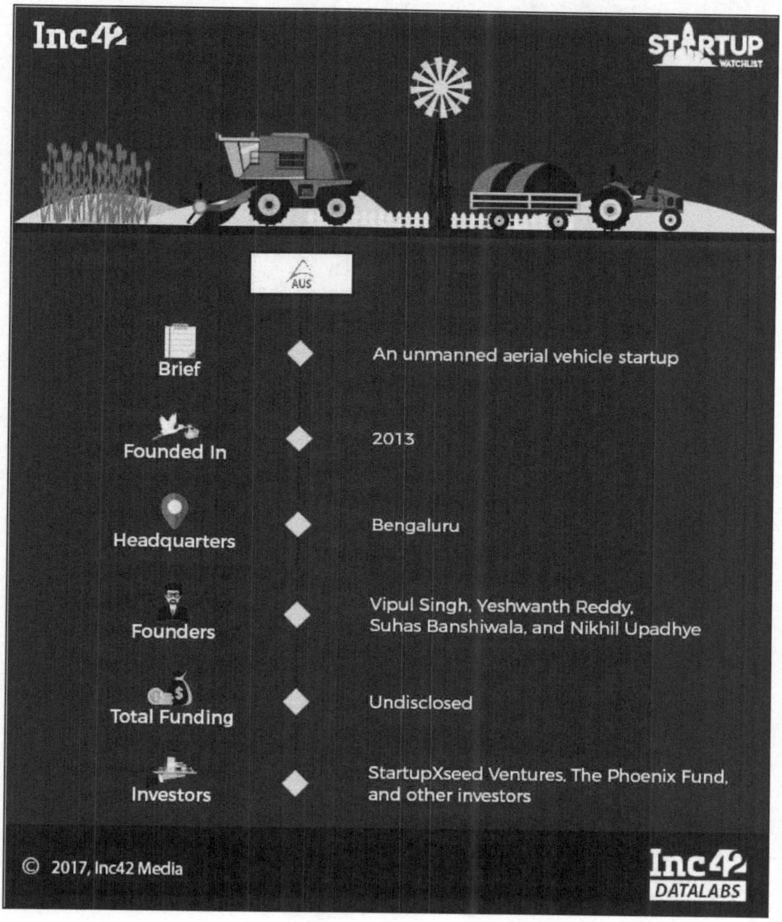

Image Courtesy Aarav Unmanned Systems Fact Sheet.

2. Intello Labs

The Bengaluru based agritech startup offers a product DIGITAL AGRI which uses computer vision algorithms to see the minutest detail on every plant and harness human intelligence to grade agricultural commodities. Intello Labs aims to revolutionize agriculture by enabling farmers, traders, millers, retailers and end users in communication though images for product quality, infestation, plant health or even soil conditions. Intello Labs uses Artificial Intelligence and deep learning for crop quality parameters like crop infestation, nutrient deficiency, harvest quality evaluation of fruits, vegetables, grains and other crops and farm to fork commodities *etc.* The startup was also recognised as one of India's Most Innovative Top 50 Emerging Software Product Companies by Nasscom in 2018.

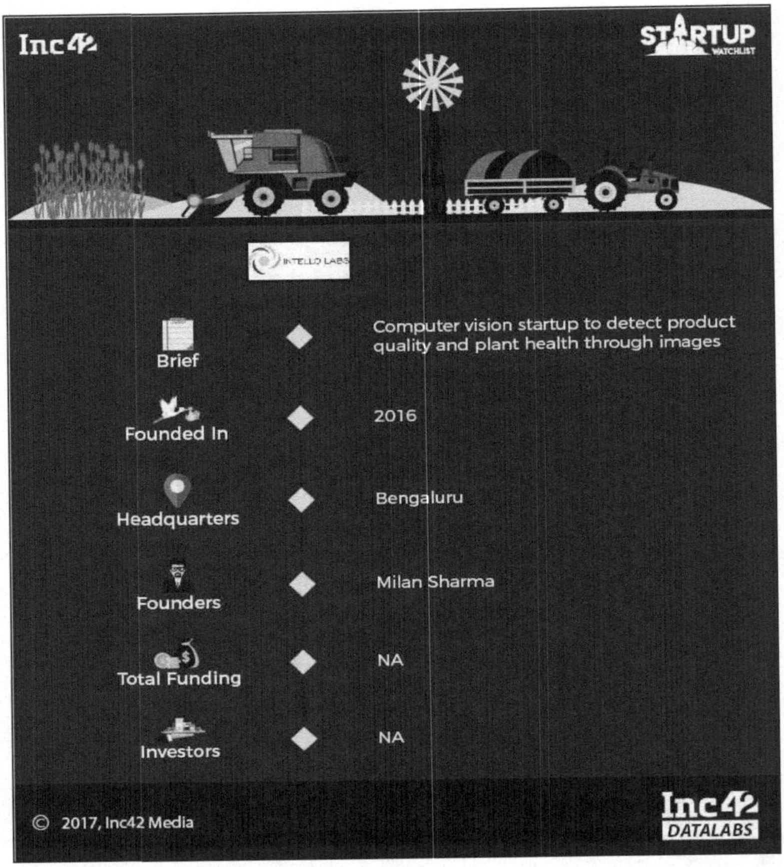

Image Courtesy Intello Labs Fact Sheet.

The startup use satellite and drone images for yield estimation, and water requirements. Farmers, procurers, field agents or quality professionals can just click an image of the crop or harvested produce and Intello Labs can provide precise insights from the images. Thus it has immense potential in digital agriculture in the country, in eliminating wastage, in food grain quality grading, and in reducing crop losses due to diseases. AI can enable farmers to switch to smarter techniques to control crop infestation and reduce human interference and this is where Intello Labs' strength lies.

3. Agribotix – Drones

Colorado-based Agribotix reportedly takes agricultural data captured by drones and conducts analyses using cloud-based software to help clients increase crop yields and profits. According to the Agribotix website, the company claims that it has experience with more than 44 crops and its clientele base spans over 45 countries. The one minute video below provides a demonstration of a drone in action:

In one case study, Agribotix claims its technology helped a soybean grower prevent damage to its crops from weeds and avoided a 13 per cent crop loss. The reported ROI amounted to $7,222.00, which included savings from crop loss avoidance and a precise herbicide application.

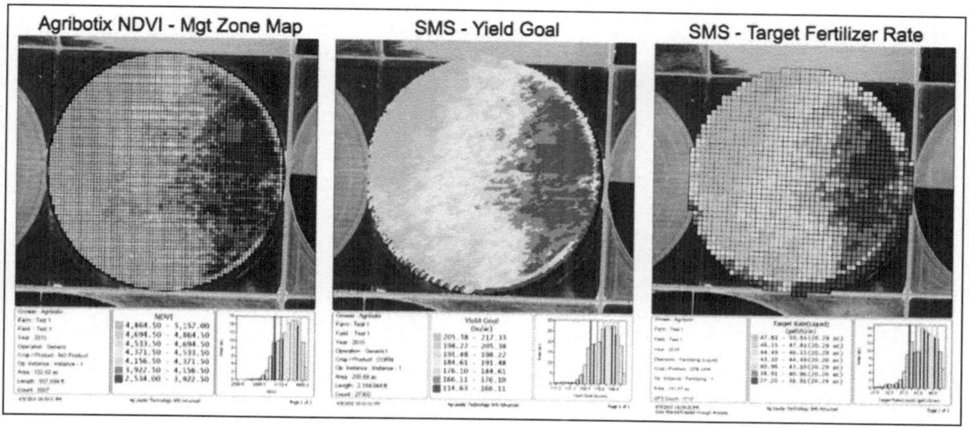

Agribotix – Drones.

In May 2017, Agribotix announced its involvement in a partnership with The Climate Corporation to expand the Corporation's suite of digital tools by integrating the ability to capture high resolution aerial images.

For example, as depicted in the image below from the company's website, these aerial images provide a contrasting view of the terrain, highlighting which areas are healthy and which require attention. Theoretically, the information derived from these images would allow more efficient budgeting and planning of farming and harvesting procedures.

Chapter 20

18 Robots Revolutionizing the Agricultural Landscape

Here are 20 examples of popular agricultural robots that carry out various functions related to farming.

1. Tertill

Tertill, a 4-wheeled robot, removes weeds independently using a nylon string under it. It is suitable for home gardens and small farms. It has a solar panel incorporated and only weighs 1.1 kg.

Franklin Robots.

2. Octinion

R and D company Octinion's strawberry harvesting robot is slated to officially debut in 2018. In comparison to Shibiya Seiki's strawberry picking robot, Octinion emphasizes the ability of its robot to prevent damage to crops. The prototype was reportedly developed two years prior as demonstrated in the video below:

Like the Shibiya Seiki strawberry robot in the previous example, Octinion's application seems to operate within a special farm layout with hanging fruit and neat rows at a specific picking level. We might imagine that fruit at various levels (or in different layouts) might be more challenging for the machine.

According to its website, Octinion claims its autonomous robot is developed with 3D computer vision which gives it the ability to only pick strawberries that it won't bruise; this equates to picking an estimated 70 per cent of all ripe strawberries.

The company reports that human labor results in 16 kg of strawberries picked and sorted per hour (with about 56 strawberries per kg), equating to about one strawberry every four seconds. Comparatively, Octonion claims that its robot "has the capacity to pick one strawberry every three seconds." This speed seems rather exaggerated, and it seems obvious that many more iterations and improvements will be required before this machine reaches human skill in picking and sorting.

Octinion does not presently feature any case studies or real-world applications of it's technologies on it's website, nor is there evidence of current paying customers (though this information might be kept secret).

The Rubion model robot from Octinion identifies and plucks ripe strawberries without bruising them. It also weighs and places them safely in the punnet.

Octinion.

3. SmartCore

SmartCore helps in collecting soil samples intelligently for analysis. It makes sure to collect them from the same point every season to analyze the field's evolution accurately.

Smart Core.

4 Fendt

Fendt's robots work in swarms across the field to plant seeds. It uses satellite-based navigation to accomplish the task.

Fendt Xaver.

5. Dino

Dino from Naio Technologies works autonomously using the GPS guidance system for weeding large fields. It can work on predefined schedules like weeding every week.

Dino Robot.

6. Agerris

The SwagBot robot from Agerris will work based on AI software control and can weed, monitor crops, and even watch cattle, eliminating watch dogs.

SwagBot Robot.

7. Mamut

Mamut uses sophisticated technologies like an AI system, compass, and six 360 degree cameras for its work. It roams the field and captures images, helping to identify diseased crops.

Mamut Robot.

8. FarmWise

FarmWise is a versatile robot used for weeding and adapts to different environments. It works with an AI-controlled mechanism and Deep-learning algorithms.

FarmWise.

9. EcoRobotix

Ecorobotix is a scan-and-spray robot targeting its enemy accurately with the help of a camera system and AI-control. It sprays 20 times less pesticide than traditional spraying methods.

Ecorobotix.

10. Thorvald

Thorvald is envisaged as a versatile robot performing many functions and replaces human labor to a significant extent. It navigates the field without damaging plants and can even carry loads.

Saga Robotics.

11. Energid

Energid is a harvesting robot mostly helping in picking citrus fruits. It can pluck one fruit in 2-3 seconds, and the overall cost is cheaper than human labor.

Energid Robot.

12. Agrobot E-Series

The E-Series is another harvesting robot with an AI system and 24 robotic arms. It not only picks strawberries but can also identify the ripe ones accurately.

13. Blue River

The LettuceBot2 robot from Blue River helps farmers in thinning out the lettuce fields. It attaches to a tractor to do its work. Herbicide resistance has become a

Blue River Robot.

primary concern for stakeholders in the agricultural industry. The increasing use of herbicides has contributed to herbicide resistance which has been documented in 250 species of weeds. The Weed Science Society of America recently concluded that herbicide resistant weeds have been responsible for approximately $43 billion worth of financial losses for American farmers.

In response to the continued challenges that weeds are presenting to farmers, Blue River Technology has launched a weed spraying machine. The *See and Spray* robot has been marketed as a safer solution to herbicide resistance with claims that it significantly reduces the the crops' exposure to chemicals. In September 2017, noted agricultural company John Deere announced its acquisition of Blue River Technology for a reported $305 million.

Blue River Robot.

Using computer vision and artificial intelligence, these smart machines can detect, identify, and make management decisions about plants in the field. The machines see every plant and determine the appropriate treatment for each. The machine can distinguish subtle differences between cotton plants and weeds of many species and sizes. The robot's nozzles target unwanted weeds in real time as the machine passes. It applies herbicide only to weeds, avoiding application on crop or areas without weeds. The machine then assesses the applied herbicide, makes adjustments, and learns as it goes.

14. Agribotix

Agribotix by AgEagle Aerial Systems is a drone robot that helps monitor the crops and even measures their health. It captures aerial photos and records videos and also has an infrared senor system.

AgEagle Aerial Drone Robot.

15. Vision Robotics

The Vision Robotics range of robots can automatically thin plants, especially lettuce, and prune vineyards. It is sophisticated with its AI-powered mechanisms.

Vision Robot.

16. RoBoPlant

RoBoPlant helps in planting seeds using its semi-automatic and fully automatic modes. It finds optimal patterns for planting and is suitable for greenhouse management.

RoBoPlant.

17. PrecisionHawk

PrecisionHawk is a versatile drone robot that can help farmers with many sophisticated functions. It leverages multispectral, hyperspectral, and LiDAR technologies to perform tasks.

Precision Hawk Drone Robot.

18. Nexus Robotics

The award-winning R2Weed2 robot from Nexus intelligently distinguishes between weeds and crops and removes weeds efficiently. It performs environmental monitoring and soil analysis alongside too.

R2Weed2 Robot.

Chapter 21

The Agricultural Robotics Marketing in the World

Today, it's the 2nd biggest market for professional robotics and estimated to be worth \$2.75 billion. According to many researchers, the agricultural robots market should have an annual growth rate of over 20 per cent until 2022. Eventually, it will reach a market capitalization of \$13 billion. This spectacular growth will be fueled by the democratization of automated machines, by their ever-decreasing costs, but mostly by the diversification of the nature of the machines. Today, a vast majority of robotics on farms are for milking cows or indoor machines. But in the very

Tractor in a Field.

near future, robots will start working outdoors. You may have already seen some automated tractors or drones but, truth be told, these are still sparse initiatives. That being said, the day when we'll see robots roaming through farms are closer than you can imagine. For instance, we already have advanced prototypes for feeder robots, surveyance, and weeding. A future à la Interstellar with fully automated farms might just be around the corner.

According to a Markets and Markets Research report, the most important applications will be in:

1. Harvesting Management
2. Field Mapping
3. Dairy Farm Management
4. Soil Management
5. Irrigation Management
6. Pruning Management
7. Weather Tracking and Forecasting
8. Inventory Management

It's then safe to say that most of today's farming is about to be disrupted by robotics.

1. Modern Farming

Starting from the premise quoted earlier, the robot-enhanced agriculture is also a question of survival. Today's farming is far from optimal on so many levels. According to an FAO report, family-run farms produce over 75 per cent of the world's food. And about 75 per cent of all farms in the world are smaller than

Modern Irrigation, while it got Better, is Still Inefficient.

one hectare. And those small farms are way more efficient than the large ones. Contrary to the rule of thumb that large-scale productions see a decrease of inputs per unit of production, farming is actually more efficient and productive at smaller scales. Bringing robots and technology into the equation will improve all farms' productivity. If we want to tackle the challenges of feeding the ever-increasing world's population, we'll need to make sure that the food production process minimizes the inputs needed and maximizes the yields. That's something only robots can help us achieve. **The future of farming is technology and data-driven.**

Let's take for example the modern irrigation. What do you think will be most productive? Using hundreds of meters of aerial pipes like we can see today in linear or central pivot irrigation or arrays of moisture sensors spread across the field, making sure the soil has the perfect moisture levels at the perfect places for crops to grow as productive as possible (without mentioning the use of weather forecasts to adapt the irrigation schedule to ensure as little water is used as possible)? Which brings us to our first point?

2. Agricultural Robots will be so much more Efficient

The first obvious improvement brought by farming robots will be that they'll be autonomous. While people will be able to focus their mind on more interesting tasks, they will be roaming the properties, relentlessly accomplishing their mission. Whether it'll be surveyance, removing weeds and pest, caring for crops,. you can be sure that you won't have to be behind them to get the job done. Thanks to online databases and forecasts, combined with the data provided by external and internal sensors, they will have a complete understanding of the situation, far beyond what any human could. This will allow them to make informed decisions, and obtain the optimal outcome. Furthermore, as all of these robots will be powered on batteries and electric motors, they will also have a positive impact on the energy consumption and carbon footprint of the farms. Some of them might even be

Milking is Already in Majority Done by Milking Robots.

perfectly independent thanks to solar panels on them. They'll be an important part of a greener future for mass-scale farming. Besides, this will also allow for a quieter exploitation. Thus it will also reduce the noise pollution generated by agriculture.

In a nutshell, by bringing artificial intelligence and around-the-clock activity, the robots will help optimize the yield of the farms far beyond what could humanly be possible. Maximizing the production while minimizing the costs in energy, water and time. Can you imagine greenhouses with automatic lighting and watering designed to get the best crops for the minimal energy? Or tractors relentlessly roaming the fields with drones surveying cattle day and night and rovers on a permanent weed-control duty? Or watering hectares and hectares of lands using only the optimal amount of water for each square meter? This is what tomorrow will be made of.

3. Robots on Farms will Improve Human Lives

Farm employment is far from a riskless job. It often exposes workers to pesticides, long working hours in the heat and sun, and poor sanitary conditions among other direct health risks. There are many serious health hazards affecting farmworkers. For instance, the Organic Dust Toxic Syndrome, caused by grain and silage dust, is a common illness among farmers. Less known, the exposure to excessive noise produced by machines can permanently damage workers' hearing. Furthermore, most of the work is physical and outside. Thus increasing the risk of accidents caused by fatigue, difficult environment and weather, and aging tools. However, only some of these risks are being covered by employment and health insurances. Some are not even covered by national legislation and regulations. By bringing robots to do these tasks, we will be reducing the risks on the human workers. Which will not only have a positive impact on their lives, but also on the people around them, and the society at large?

4. Robots will be Able to Farm 24/7 - No more Driving Tractors in the Middle of the Night

And obviously, as mentioned above, the more efficient the farming, the better we all are. Maximizing yield while reducing the environmental and societal costs can help us tackle the challenges of a growing population and the lack of efficiency of modern intensive farming. We can all agree that a healthier planet would be a very strong improvement in everybody's lives!

5. There are many Questions Raised by the Farming Robots

And they might impact their democratization more than you could suspect. As with any technological revolution, many questions need to be answered. Today, the focus of governing entities is on the legislation and regulation of the potential agricultural robots. Due to the fact that these robots will be in an open environment and will probably be massive pieces of machinery. While it's for obvious security reasons (you don't want a pilotless combine to go on a road trip), it can create serious entry barriers to startup trying to break into this field.

Indeed, the certification process can already decide of the fate of a prototype. For instance, the Centéol developed by AgreenCulture is a promising autonomous robot weighing 300kg. However, until it gets its certification of conformity with the European Union norms, it can't be officially released on the market. This is just one of many examples of speedbumps that the farming robots will have overcome to become mainstream.

6. Governments are Jumping on the Bandwagon for Agricultural Robotics

Seen all that legal blur and balancing it against the numerous advantages, some governments are starting to realize the possible impact and the need for action. Whether it's the positive fallback of leading the way on that matter or the risk of missing the train on that new revolution. For instance, France, one of the largest agricultural production in the world, through its national agronomic and research agency IRSTEA, has announced in November 2017 the creation of an accelerator program "RobAgri". Its goal couldn't be any clearer: *"Intensify the conception, validation, and dissemination of tomorrow's robots"*.

7. Investing in Agricultural Robotics is Sowing the Seeds of a Prosperous Future

One of the biggest government programs on that topic in the world, RobAgri gathers almost 50 public and private stakeholders in order to impulse a dynamic collective to conceive robotic systems fitting the needs of farmers and markets. Besides its workgroup, RobAgri is partnering with Naio Technologies to organize a yearly convention about agricultural robotics. The Forum International de la Robotique Agricole (FIRA) will hold its 3rd edition on December 13th and 14th in Toulouse, France. Likewise, back in 2014, the UK set out an "agri-tech" strategy with over £150M in public funding to develop and commercialize agricultural technologies.

8. You too can have your own Agricultural Robot

The robotic farming isn't reserved to only big exploitations. If you wanna be part of the agricultural revolution brought by robotics, we can only suggest you get yourself a FarmBot. This open-source project allows you to build your own automated veggie garden. Powered by a Raspberry Pi and an Arduino board, it offers to bring a CNC head to your raised bed. Thanks to its open-source software you can access your crop from anywhere and make sure everything's running alright. Thanks to its link with OpenFarm.cc and weather data, FarmBot can optimize the care of your crops. According to its manufacturer, it can save you up to $1,500 a year for your vegetable consumption. Or if you're looking for a smaller-scale project, there's the MoAgriS project over at Hackaday that will be a great start.

If you're looking to automate your garden's irrigation and watering systems, then look no further than the GreenIQ Smart Garden Hub available at RobotShop.com. Using data from weather services or from moisture sensors (3 are included with the base product), this hub can save you up to 50 per cent of your water consumption. You can use its additional master channels for a water pump and a fertilizer tank. It can also take control of your lighting based on sunset and sunrise schedules. And obviously, you can control it from your computer or your smartphone from anywhere in the world. Or ask Alexa to do it for you! (Jpvalery, 2018).

Chapter 22

Enabling Technologies for Future Robotic Agriculture Systems

Future Robotics Agriculture Systems.

A wide range of technologies will enable the transition of agricultural robotics into the field. Some technologies will need to be developed specifically for agriculture, while other technologies already developed for other areas could be adapted to the agricultural domain for example, autonomous vehicles, artificial intelligence and machine vision. Here we briefly review the current status, opportunities and benefits of various enabling technologies from hardware to software, multi-robot systems and human robot systems.

1. Vision Robotics – Planting and Seeding

Vision Robotics' technology reportedly integrates algorithms with sensor technology to bring automation to lettuce farming and vineyards. Specifically, computer vision allows robots to generate 3D maps and models of areas of interest and then to complete various tasks within those parameters. For example, "thinning" is a process in farming where seeds are adequately spaced apart during planting to allow for optimal crop growth. It can also be a time-consuming process. Vision Robotics' automates the lettuce thinning and results are shown in the example image below:

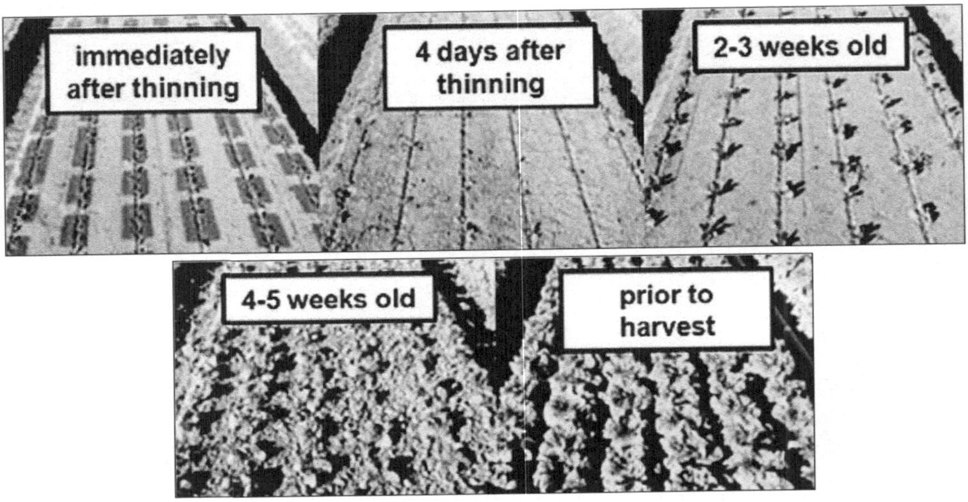

Vision Robotics.

Future initiatives appear to include the development of precision weed removal technology using herbicides and the company is actively seeking strategic partners. Vision Robotics has not published an anticipated timeline for this effort.

Machine vision approaches offer significant opportunities for enabling autonomy of robotic systems in food production. Vision-based tasks for crop monitoring include phenotyping (Pound *et al.,* 2017) classifying when individual plants are ready for harvest (Kusumam *et al.,* 2017), and quality analysis (Barnes *et al.,* 2010), *e.g.* detecting the onset of diseases, all with high throughput data. Vision systems are also required for detection, segmentation, classification and tracking of objects such as fruits, plants, livestock, people, *etc.*, and semantic segmentation of crops versus weeds (Haug *et al.,* 2014, Lottes *et al.,* 2016 and Bosilj *et al.,* 2018) *etc.* to enable scene analysis (understanding "what" is "where" and "when") and safe operation of robotic systems in the field. Robotic vision in agriculture requires robustness to changes in illumination, weather conditions, image background and object appearance, *e.g.* as plants grow, while ensuring sufficient accuracy and real-time performance to support on-board decision making and vision-guided control of robotic systems. Active vision approaches, integrating next-best view planning, may be needed to ensure that all the relevant information is available for robotic decision-making and control, *e.g.* where the fruit or harvestable part of a crop is occluded by leaves or weeds. Approaches based on analysis of 3D point clouds, *e.g.* derived from stereo imagery or RGB-D cameras, offer significant promise to achieve robust perception in challenging agricultural environments (Kusumam *et al.*, 2017 and Smith *et al.*, 2018). Machine vision is already making an early impact in animal monitoring, *e.g.* for weight estimation, body condition monitoring (Hansen *et al.*, 2018) and illness detection (Jabbar *et al.*, 2017) in pigs, cattle and poultry. Individual animal identification, *e.g.* using facial recognition techniques adapted from work in human facial biometrics (Hansen *et al.*, 2018) will allow more targeted precision care and timely interventions for individual animals, thereby ensuring their healthcare and wellbeing as well as optimizing farm production.

Vision Drone.

Robotic vision often depends closely on machine learning from real-world datasets, with approaches such as deep neural networks (Sa, *et al.*, 2016, Pound *et al.*, 2017 and Sørensen and Bochtis, 2010) gaining traction and further raising the possibility for robots to share their knowledge by learning from Big Data. An open challenge in robotic vision and machine perception for robotic agriculture is to enable open-ended learning, facilitating adaptation to seasonal changes, new emerging diseases and pests, new crop varieties, *etc.* Most existing work considers only the initial training phase prior to deployment of a robot vision system, but not the ongoing adaptation of the learned models during long-term operation. The development of user interfaces for "ground truthing" and semi-supervised learning in robotic vision systems for agriculture is also an open challenge.

Discussed about the machine vision applications in agriculture is mainly three type's non-destructive measurements, visual navigation, and behavioral surveillance. "Fixed forward field of view" is important in Machine vision system. There are three FOV (Field of View), far FOV, near FOV, lateral FOV. FOV gives accurate guidance for agricultural robots. Fixed looking cameras have limitations. FOV of high pixels is used in row guidance. Instead of using one type of FOV, all three types of FOV is combined and variable FOV is made so that three types of views can be obtained in one robot. Variable FOV camera arrangement was used to guide an agricultural robotic platform which can be viewed using a laptop to drive wheels. Machine vision based row guidance is to enable stable navigation of robot, distinguish green corn plants and soil background. Size of the robot is optimal. GPS is attached to the robot to find location. Morphological operations were used to calculate guidance lines in the field. Far FOV is more efficient compared to near FOV (Xue *et al.*, 2012).

Machine Vision System.

2. Shibiya Seiki – Strawberry Harvesting

In 2013, Japanese company Shibiya Seiki made headlines with the announcement of a $50,000 robot reportedly capable of picking strawberries. Results from a research paper published by the company reported that the machine showed "potential for practical use" – and is clearly not in use with businesses today. It has been estimated that it takes a human 10 to 15 minutes to pick a quart of ripe strawberries, and – this robot seems rather far from beating human performance in the near term.

The greenhouse setup for Shibiya Seiki's research was customized, complete with a hanging strawberry garden setup with a particular density of fruit and a particular set of distances between plant beds. While the machine likely would be of little use in a normal (read: un-instrumented) berry farm setting, this application may help generate ideas about the super-efficient farm setups of the future (optimized for machine performance, not human performance). The robot was developed in collaboration with Japan's National Agriculture Food Research Organization (NARO). NARO's current research efforts appear to include the development of fully-automated farms. It is unclear exactly when the organization expects to realize its ambitious effort.

3. nFF Robotics – Apple Harvesting

Historically, apple harvesting has been conducted manually. In 2012, an estimated 4.2 million apples were picked in the U.S. Israel-based FF Robotics has reportedly been working to perfect an automated apple-picking robot as demonstrated. While the company showcases instances of media coverage on its website, it is unclear what clients have invested in the technology to date.

4. Robotic Platforms

Agricultural platforms can be divided into domain- and task-specific robots designed to perform a specific task on a given crop in a pre-defined domain, and generic platforms designed to perform several tasks in different domains. Both are likely to play important roles. Since farms in general have very different infrastructure, early robots may be able to operate only on a given farm and only to a limited extent across different farms. Similarly to current farm vehicles, we may see therefore a combination of robots adapted to a specific task and the emergence of multi-purpose robots able to carry out a multitude of different tasks, analogous to the myriad use cases of the modern tractor.

A common challenge is that most current robotic platforms are not robust to real-world conditions such as mud, rain, fog, low and high temperatures. For example, most current manipulators are not equipped to deal with humidity in glasshouses. The development of rapid prototyping techniques and lowcost processors have led to an explosion in the use of 3D printing and "maker" technology, raising the potential of lowcost robotic platforms for a variety of applications. The use of embedded software enables highly configurable and application-specific platforms that can use common hardware modules and be adapted to a variety of roles. While such approaches have been used extensively in UAVs and smaller-scale robots, there is much scope for the expansion of robotics in Agri-Food on a much wider scale. Issues that need to be addressed to migrate from prototypes to robust commercial platforms include robustness and reliability, power management (the platforms need to be able to operate all day, in some cases 24/7, for extended periods), usability (the platforms must be able to be used effectively by non-specialists), maintenance (*e.g.* selfdiagnosis) and integration with mobile communications. Further challenges include better characterisation of the mechanical properties of soil relevant to these robots, ruggedised platforms capable of operating in inclement weather, real-time sensing and control algorithms to adapt locomotion strategies to an ever-changing environment, and co-design of locomotion with other capabilities. For example, how does crop/fruit collection affect locomotion? What locomotion capability do we need to enable efficient sensing of crops?

5. Locomotion

Agricultural robots need to move in challenging dynamic and semi-structured environments. Ground robots needs to travel on uneven, inhomogeneous, muddy soil, while aerial vehicles need to operate for long periods of time, in different weather conditions. Current agri-robots are mainly designed by borrowing technology from other sectors (*e.g.* drones) or as an add-on to existing platforms (*e.g.* autonomous tractors). As such, they may be not fully optimised for their tasks, or may retain some of the limitations of existing platforms. UAVs can fly using multiple rotors or a fixed wing platform (with precision of location in the former and extended flight time in the latter), whereas ground platforms need to be able to locomote on rails and concrete floor in greenhouses, on gravel or grass in polytunnels, and in extremely muddy and difficult terrain in open fields (Grimstad and From, 2017).

Locomotion in Robotics.

We will therefore see a wide variety of robots being developed with different means to locomote. Compared to tractors, these robots are extremely lightweight, but as robots (or autonomous tractors) are to perform more energy-demanding task, the robots will also increase in size and weight. Most agricultural robots today run on batteries and electrical motors. Future developments will depend on how the battery technology evolves, but we will probably see both electric and combustion engines in the field for the foreseeable future. A key aspect of any robotic platform is the impact of the weight and locomotion system on the ground and crops, and therefore different platforms have been used, including tracked and multiple wheeled robots. The platforms are also dependent on the required task, for example, heavy crop harvesting (such as volume arable or root vegetables) will need a heavier platform than soft fruit picking. Legged robots have the potential of minimising their footprint, while maximising the flexibility of locomotion (*e.g.* moving sideways or in narrow spaces between crops, *etc.*). Their agility, combined with the possibility of carrying specialised sensors, may help unlock the full potential of precision agriculture.

6. Manipulators

Manipulators will be needed for a range of tasks in future agriculture, replacing dexterous human labour, reducing costs and increasing quality, or performing operations more selectively than current larger machinery like slaughter harvesters. Work in this direction is ongoing, with soft grippers used for experimental work on selectively harvesting mushrooms, sweet peppers, tomatoes, raspberries and strawberries. Other applications such as broccoli harvesting can be performed with cutting tools, but will also require gentle handling and storage of the picked crop. In the open field, and for protected crops, there are complementary tasks to harvesting where manipulators can also play an important role. This includes mechanical weeding, precision spraying, and other forms of inspection and treatment. Manipulators will also be needed for the increased automation seen in food handling applications, such as large automated warehouses.

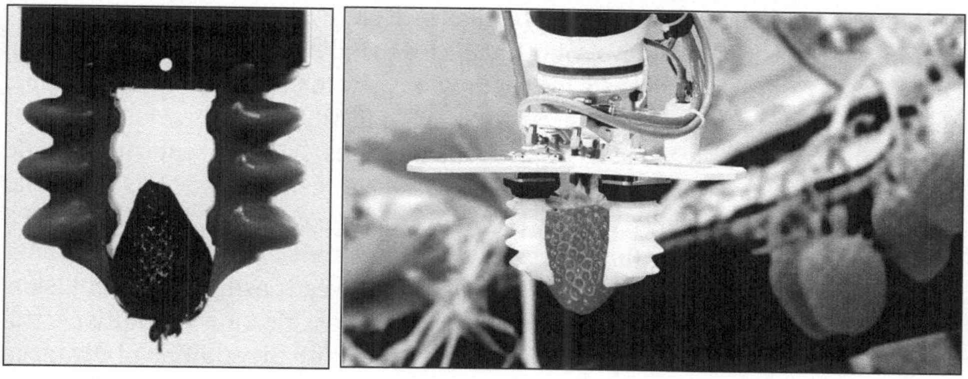

Manipulators for Picking Strawberry.

7. Sensing and Perception

The integration of sensor systems within autonomous robotic systems offers the significant potential for new measurements that would otherwise be unobtainable. For example, current work addresses large area field mapping for bulk moisture by mobile robots, through the application of cosmic ray sensors adapted from the static COSMOS approaches (STFC Research Grant ST/N006836/1, 2016). Significant advances in satellite or drone based remote sensing capabilities open opportunities in monitoring crop growth status with unprecedented temporal and spatial resolutions while at an affordable cost. Many open source satellite datasets (*e.g.* from the European Space Agency, 2018) are freely available for farmers. Robotic platforms further offer the possibility of forensic testing of soil with the geotagging and immediate results from sampling sensors (such as laser-induced breakdown spectroscopy), or secure collection of samples for later analysis in a systematic and uncontaminated manner. The use of compact robots and on-board secure collection systems will further enable a step change in the regulatory efficiency and reliability of land management systems using robots.

8. Localisation and Mapping

The use of GPS navigation in agriculture has become almost ubiquitous with the deployment of RTK allowing accuracy of centimetres for the automated positioning of large farm machinery such as tractors and combined harvesters. More recently, approaches using data manipulation of the GPS signal alone has shown promise to deliver equivalent accuracy without the cost of extra radio beacons. Accurate location data is not confined to unmanned vehicles with GPS, as precise localisation systems are available using visual fiducial markers and/or optical, acoustic or radio beacons, depending on the speed and accuracy required.

Sensor information is also required in detecting objects and risks in field in order to ensure safe operation of robotic vehicles. To minimise damage to crops, the accuracy of relative positioning and navigation is more important than that of absolute navigation and position as provided by RTK GPS in many applications. For example, it is desirable to drive robotic vehicles to follow crop lines in accuracy of centimetres or follow the tracks left by previous tractor operations. Multi-modal systems based on a combination of GPS, INS, LiDAR, vision, *etc.* have further potential for providing accurate and robust solutions, without requiring in-field infrastructure such as beacons. Several attempts have been made to utilise seed and weed mapping concepts by passively recording their geospatial location using RTK GPS. Farming robots can be further equipped with pattern classification techniques that can predict the density and species of different weeds using computer vision. Other methods focus on a dense semantic weed classification in multispectral images captured by UAVs (Zhang and Kovacs, 2012). With the addition of advanced vision systems, including depth perception, scanning sensors such as LiDAR and artificial intelligence for decision making and classification, the concept of precision can be taken to another level. The ability offered by ground based robots to precisely control the location of scanning sensors, such as LiDAR, opens up the possibility to return quantified biomass estimates over whole crops as well as related phenotypic data, such as growth rates and morphology, through the integration of accurate location data with rangefinder scans using Simultaneous Localisation and Mapping (SLAM) techniques. Similarly, robotic sensing platforms offer the potential for broad area analysis of insect pest or pollinator movement and their speciation, utilising 3D microphones alone or in combination with light backscatter measurements to enable daylight measurements of characteristic flight trajectories. Thematic maps can be built up for diseases, pests or weeds, which enable variable rate treatments, a key concept in precision agriculture.

9. Crop Monitoring

The use of both land-based and aerial platforms can allow the third dimension to be accurately added to the management of crops using data fusion and SLAM techniques. This can be combined with virtual reality or augmented reality (VR/AR) systems to provide monitoring and intervention possibilities to an individual plant scale. Long-term data collection will further enable the modeling of crops over time, for example, tracking the development of the crop canopy and thus improved prediction of future growth patterns. Such ground and aerial robotic platforms offer additional prospects for enabling localised extremely high signal to noise, high resolution sensing that may not be achieved by passive remote (satellite) or semi-remote (rotary or fixed wing UAV) sensing technologies. At the simplest level, these robotic platforms offer the potential to extract close proximity (within 10s of millimetres) reflectance and transmission. Multispectral Imaging (MSI) data helps compensate for the erroneous measures that occur due to the surface topology and orientation of individual crop tissues. At a more advanced level, the use of robotic manipulators to locate sensors around crops or livestock could enable responses to be tested and examined, through applying artificial stimuli. For example, through applying a focused beam of light at specific areas of crop tissue, and then modulating the spectrum and intensity, it is possible to drive photochemistries within specific parts of plants, *e.g.* stems, young leaves, senesced older leaves, *etc.*, which can then be sensed via multispectral imaging. In this way significantly greater phenotype information may be recovered from across plants than could be achieved by passive fixed imaging detectors alone. Similarly, the cell structures and arrangements within fruits, vegetable and meats may be non-destructively examined in high resolution, *e.g.* for mapping subcutaneous bruising in fruits or fat ratios in meats. Nutrient and water stress of crops can be assessed by fusing MSI data with other data sources. Combining these assessments with crop growth models gives a better prediction of yield and loss, which leads to improved farming management and better food supply chain management.

10. Planning and Coordination

The true potential of robotics in agriculture will be harnessed when different types of robots and autonomous systems are brought together in a systemic approach. For example, UAVs are an excellent platform for environment monitoring, but with limited payloads and operational durability they are constrained when it comes to delivery of intervention or treatments on a larger scale. Hence, ground and airborne vehicles need to be integrated into heterogeneous fleets, coordinated either centrally or in a distributed fashion. Planning, scheduling and coordination are fundamental to the control of multi-robot systems on the farm, and more generally for increasing the level of automation in agriculture and farming. For example, intelligent irrigation systems can respond to the change of weather conditions and crop growth status to automatically optimise the irrigation strategy so as to reduce the use of fresh water without loss of yields. The optimised strategy (*e.g.* when, where and the amount of water) is then implemented by computer-controlled irrigation equipment. Such coordinated fleets also pose requirements for in-field communication infrastructure, such as Wi-Fi meshes, WiMAX ad-hoc networks, 5G approaches or other proprietary peer-to-peer communication methods deployed in field. On a larger scale, the heterogeneous fleets' deployed in-field can also include collaborating humans sharing the working environment with their robotic counterparts, giving rise to interaction and communication requirements between the robots and the human operator and workers in this context. Example applications include in-field logistics, where vehicles need to be scheduled for area coverage and routing problems. More generally, holistic approaches to fleet management are required which fully integrate component methods for goal allocation, motion planning, coordination and control (Sørensen and Bochtis, 2010). These sub-problems have so far largely been studied in isolation, so basic research on integration and scaling to real-world scenarios is required. Aspects of swarm robotics could potentially be applied to fleet management systems in agriculture, as in the EU-funded ECHORD++ projects SAGA and MARS (Albani *et al.,* 2017). To enable robot-human collaboration, the fleets also need to be aware of the presence of humans and to predict likely human actions in order to anticipate potential collisions and ensure safety. In return, the motion of robotic systems needs to be legible to humans, to facilitate acceptance by and cooperation with their human counterparts.

11. Manipulation

Automated manipulation and grasping of food items presents a series of unique challenges compared to other sectors. These include significant natural size and shape variations between examples of the same product, heterogeneous positioning of products (*e.g.* during harvesting) and the fragile nature of food products. Some areas of food harvesting have been successfully automated but these solutions are best suited to situations where the entire content of a field becomes ready for harvest at the same time, *e.g.* grains or root vegetables. If plants fruit over an extended period of time with only some ready to harvest at any particular time (*e.g.* tomatoes) automation struggles. This is because discrete items must be harvested individually without disturbing those around them and, due to the dexterity, advanced perception and decision-making to required human labour is still widely used. Soft robotics (Shepherd *et al.,* 2011 and Kim *et al.,* 2013) is expected to play an important role. Soft end-effectors and grippers are needed for gently handling soft fruit and vegetables, such as soft robot hands for lettuce harvesting and suction devices for picking apples. Robots are increasingly made softer also on the actuator/joint level. Whereas stiff robot arms are suitable for blind operation in a factory environment, an agricultural manipulator requires sensorimotor coordination to achieve its task. Some tasks also require the right amount of force to be applied, dictating a force-based rather than position based approach to control. In general, grasping and manipulation applications in Agri-Food require robustness to the unforeseen, while maintaining their ability to actuate with precision. One way to achieve this is through variable stiffness actuators (Abeach *et al.,* 2017) which incorporate elastic structures, much like humans. The development of compliant manipulators and grippers will in turn transform and simplify the design of agricultural robots by reducing the need for complex visual and tactile sensors. For this potential to be fully unlocked, novel design and control techniques need to be developed. Grasp planning is also a significant challenge. The most common approach is to use vision systems to locate products and use this to direct the grasp. However, this approach can fail if the object to be grasped is partially obscured by other products or foliage. Vision alone provides only limited data about an object during grasping and picking; human operators also use tactile feedback to adjust their action as a product is grasped to ensure it is picked successfully.

Vision System.

a) Dual-Arm Manipulators

The idea of using two manipulators with shared workspaces was first developed in industry for the assembly of parts and components. Currently, that idea has been renewed, and such a device has been called a dual-arm manipulator. Robotics companies are introducing products in the industrial market (ABB YuMi, 2018), but research is ongoing for applications in the agricultural sector (Bechar and Vigneault, 2016). The advantages of using dual-arm manipulators can be found for (1) bimanual grasping, (2) grasping with one arm and cutting with the other arm, (3) handling deformable products, (4) moving away leaves and branches with one arm to obtain products with the other arm, *etc.* (see Figure 8).

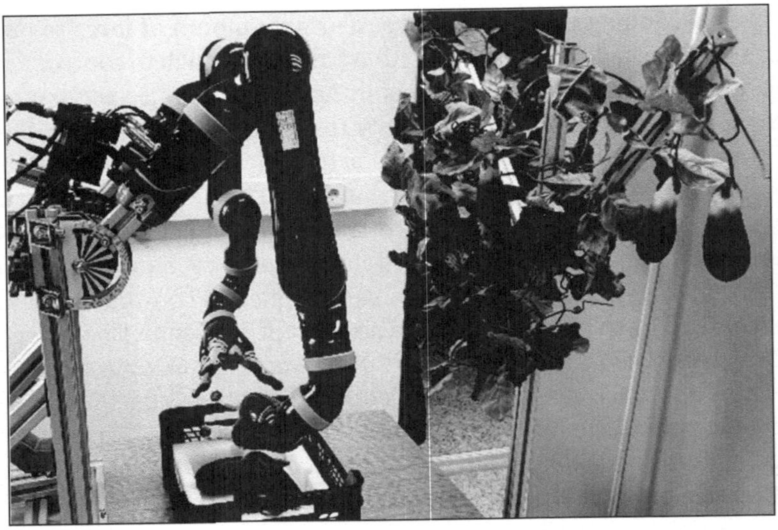

Figure 8. Dual-arm Manipulator Harvesting Eggplants.

b) Unmanned Ground Robots for Agriculture

Ground autonomous robots were based on the deployment of new technologies fostered, in the beginning, for military applications and demonstrated important achievements to adapt military vehicles as autonomous robots. When those technologies were made commercially available, especially GPS, they were used to convert conventional vehicles into ground autonomous vehicles and, then, autonomous tractors for agriculture applications emerged. After demonstrating the capabilities of these agricultural robots, the developers focused on the design of agricultural vehicles following robotics criteria.

c) Approach to UGVs (Unmanned Ground Vehicles) for Agriculture

The commercial availability of the Global Navigation Satellite System (GNSS) has provided easy ways to configure autonomous vehicles or navigation systems to assist drivers in outdoor environments, especially in agriculture, where many highly accurate vehicle steering systems have become available (Autopilot, 2020 and AutoTrac, 2020). These systems aid operators in the precise guidance of tractors using LIDAR (light/laser detection and ranging) or GNSS technology but do not endow a vehicle or tool with any level of autonomy. Other critical technologies, such as safety systems responsible for detecting obstacles in the robots' path and safeguarding humans and animals in the robots' surroundings as well as preventing collisions with obstacles or other robots, must be incorporated to configure autonomous vehicles. Additionally, wireless communication between the robot and the operator and external servers (cloud, CPSs, and IoT technologies) will be critical to incorporate decision-making systems built on Big Data analysis as well as to expand decision processes into machine learning and artificial intelligence fields.

The technology required to deploy more robotic systems into agriculture is available today, as are the clear economic and environmental benefits of doing so. However, although agricultural machinery manufacturers have not missed the marketing potential of showing concepts (New Holland, 2020 and CASE, 2017), these manufacturers are unwilling to deploy fully robotic systems into the market. Therefore, additional hard work must be made by both researchers and private companies to discover new solutions, as encouraged by the Standing Committee on Agricultural Research (EU SCAR, 2016). In addition, farmers are normally reluctant to use new technologies and thus emulation tools for new equipment will help in introducing agricultural robotic systems (Tsolakis *et al.,* 2019). Autonomous systems used for precision agriculture tasks involve the coordination of the devices as explained in Figure 9.

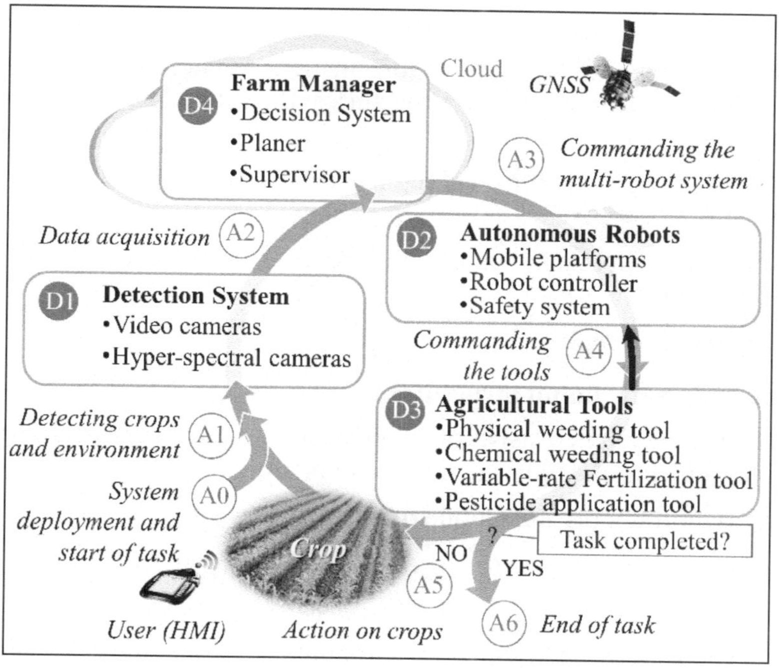

Figure 9. Components of a Precision Agriculture Robotic System.

The development of ground autonomous vehicles for agriculture focused on two approaches:

(i) Automation of conventional vehicles (ii) design of specific robot structures.

i) Automation of Conventional Vehicles

The first approach consisted of retrofitting commercial agricultural tractors with (1) actuators to manage a vehicle's throttle and steering; (2) sensors to obtain information about the vehicle's environment and self-position; and (3) computers to manage the information obtained by the sensors. The relationships among these systems are illustrated in Figure 9 along with their working process, configured as a sequence of actions (Ai, i = 0 to 6). This process starts with system deployment, A0, and then the other actions are cyclically repeated until the conclusion of the agricultural task–field completion, A6. These actions are as follows:

A0 The system is deployed in the working field.

A1 The sensors onboard the multirobot system,

D1 Detect crop, soil, and environmental features in a small area in front of the robots.

A2 The collected data are sent, via the robot controllers, to the decision support system, which determines the subsequent commands for the robots and agricultural implements.

A3 Motion commands and intervention instructions are sent to the multirobot systems, which follow the planned trajectories; the robots move to the supplied positions.

A4 The robots send the pertinent commands to the agricultural tools to perform the treatment.

A5 The treatment is applied; then, the process repeats from action A1 or exits if the task is complete and

A6 The task is complete.

ii) Design of Specific Robot Structures

D1 Detection system consists of video carmeras and hyper-spectral cameras. Robotics for agriculture applies to both manipulators

D3 Agricultural tools and mobile robots

D2 Autonomous robots and

D4 Farm manager.

The second approach to configure autonomous mobile robots for agriculture relies on the design of specific robot structures. This design is performed by considering the different techniques used in the development of general ground mobile robots, which are based on wheels, tracks, and legs, as locomotion systems. Recently, some hybrid locomotion systems based on wheels and legs have appeared in the literature (wheel-legged robots) (Figure 10 a), exhibiting interesting characteristics. Independently of the locomotion structure used, a robot also has to include elements to sense, act, and control the robot's subsystems, as indicated in the description of the retrofitting of conventional vehicles (actuators, sensors, and computers).

Institutions, Year, Characteristics and Applications of Automation in Agriculture

Sl. No.	*Institutions*	*Year*	*Characteristics*	*Applications*
1	Stanford University (United States)	1996	An automatic control system was developed and tested on a large farm tractor using four GPS antennas.	(O'Connor *et al.*, 1996)
2	University of Illinois (United States)	1998	A guidance system created by integrating a sensor based on machine vision, RTK-GPS and a geometric direction sensor.	The fusion integration methodology was based on an extended Kalman filter and a two-dimensional probability density function statistical method (Noguchi *et al.*, 1998).
3	Carnegie-Mellon University (United States)–Demeter project	1999	This is a self-propelled hay harvester for agricultural operations.	The positional data are fused from a differential GPS, a wheel encoder (dead-reckoning), and gyroscopic system sensors (Pilarski *et al.*, 2002).
4	Carnegie-Mellon University (United States)–Autonomous Agricultural Spraying project	2002	The objective was to automate ground-based vehicles for pesticide spraying to achieve a system that was significantly cheaper, safer, and friendlier to the environment.	A remote operator was able to supervise the night time operation of up to four spraying vehicles (Stentz *et al.*, 2002)
5	LASMEA-CEMAGREF (France)	2001	This study investigated the possibility of achieving vehicle guiding using a carrier phase differential GPS as the only sensor.	The vehicle heading was derived according to a Kalman state reconstructor and a nonlinear velocity independent control law was designed (Thuilot *et al.*, 2001).
6	University of Florida (United States)	2006	An autonomous guidance system for use in a citrus grove was developed on the basis of machine vision and laser radar.	The system was tested on a curved path at a speed of approximately 3.1ms1 (Subramanian *et al.*, 2006).
7	University of Aarhus and theUniversity of Copenhagen (Denmark)	2008	An automatic intra-row weed control system was connected to an unmanned tractor and linked via a hydraulic side shifting frame attached to the rear three-point hitch of the vehicle (Nørremark *et al.*, 2008).	

Sl. No.	Institutions	Year	Characteristics	Applications
8	RHEA (Robot Fleets for Highly Effective Agriculture and Forestry Management) consortium (Figure 10 a)	2014	Automation and sensor integration of a fleet (three units) of medium-sized tractors that cooperated and collaborated in physical/ chemical weed control and pesticide applications for trees	(Emmi *et al.*,2014, Emmi *et al.*,2014 and Gonzalez-De-Santos *et al.*, 2016).
9	Carnegie-Mellon University (United States), 2015		Self-driving orchard vehicles for orchard tasks, *i.e.*, tree pruning and training, blossom and fruit thinning, fruit harvesting, mowing, spraying, and sensing (Bergerman *et al.*, 2015).	
10	University of Leuven (Belgium)	2015	Tractor guidance using model predictive control for yaw dynamics (Kayacan *et al.*, 2015).	

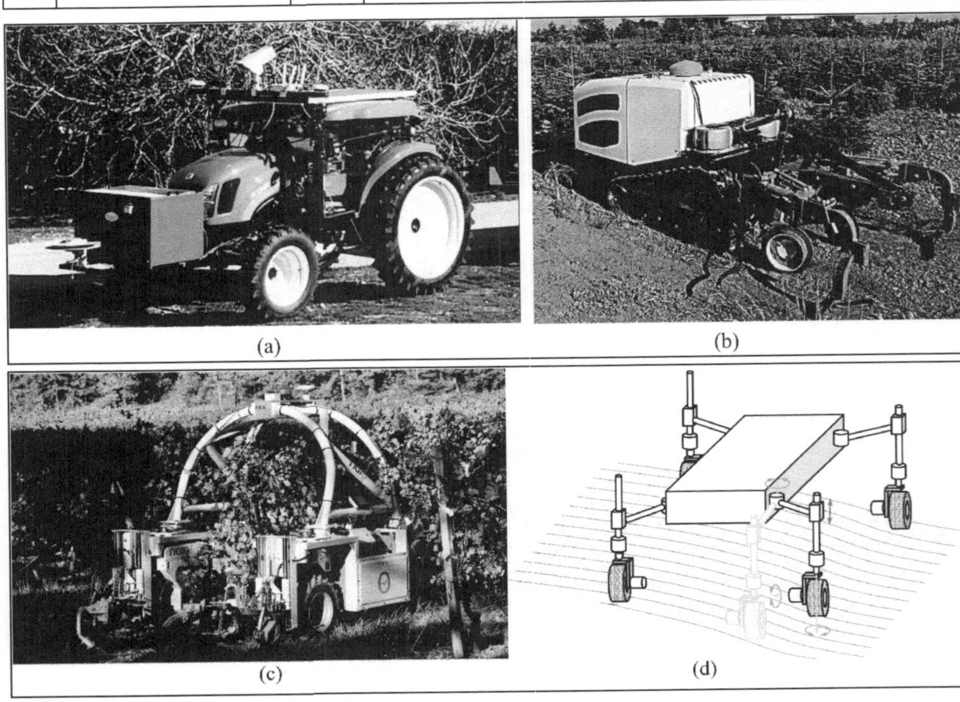

Figure 10. Mobile Platform Structures: (a) Front steering wheels and rear traction wheels (RHEA prototype); (b) Skid steering wheels (courtesy of AGREENCULTURE); (c) Independent steering wheels (Courtesy of Naïo Technologies) and (d) A wheel-legged prospective structure.

Vehicles, Year, Steering scheme and Applications of Robos in Agriculture

Sl. No.	Vehicles	Year	Steering Scheme	Applications
1	BoniRob	2009	Independent steering/ traction wheels in 1-DOF legs (wheel-legs).	General agricultural tasks– BoniRob is based on an app concept similar to the intelligent phone scheme that gives third parties the possibility to integrate their own modules for specific applications(Ruckelshausen *et al.*, 2009).
2	AgBot II	2014	Two front skid steering wheels and two rear caster wheels.	Fertilizing tasks in large-horticultural crops; weed management: detects and classifies weeds. Destroy weeds using mechanically or chemically devices (Bawden *et al.*, 2017).
3	Ladybird	2015	Steering scheme: Independent steering wheels.	Assessment of crop using hyperspectral cameras, thermal and infrared detecting systems, panoramic and stereovision cameras, LIDAR, and GPS (Underwood *et al.*, 2015).
4	Greenbot/Precision Makers	2015	Independent steering/ traction wheels.	Regularly repeated tasks in agriculture and horticulture (Precision Makers Greenbot, 2020).
5	Cäsar	2016	Independent steering/ traction wheels.	Autonomous use in enclosed fruit plantations and vineyards.Pest and soil management, fertilization, harvesting, and transport (Raussendorf, 2020).
6	RIPPA	2016	Independent steering/ traction wheels.	Vegetable growing industry; spot spraying of weeds using a directed micro-dose of liquid (Bogue, 2016).
7	Vibro Crop Robotti	2017	Skid steering wheels.	Precision seeding and mechanical row crop cleaning (Kongskilde, 2017).
8	Céeol	2019	Skid steering trucks.	Furrow preparation, fertilization, weeding, harvesting, soil analysis (AGREENCULTURE–2020) (Figure 10 b)

Sl. No.	Vehicles	Year	Steering Scheme	Applications
9	Naïo	2019	Independent steering/ traction wheels.	Mowing, leaf thinning, trimming, *etc.*, for vineyards (Naïo Technologies, 2020) (Figure 10 c)

d) Unmanned Aerial Robots for Agriculture

An aerial robot is an aircraft with no human pilot on board. This type of robot can be remotely operated by a human operator or autonomously operated under onboard computer control UAV. These devices, popularly known as drones, are built in a variety of types defined by the technology used to fly the drone, *e.g.*, fixed-wing (planes), single-rotor helicopter, hybrid system (vertical take off and landing), and multirotor. Although drones were originally developed for military equipment, drone use has rapidly expanded to other applications that include surveillance, aerial photography and video, delivery of goods, weather forecasting, traffic forecasting and agriculture, where drones are being used to support spatial data collection to influence further policies and decisions (Sylvester, 2018).

Multirotor technology is the most used in drones over the last decade. This type of vehicle is lifted and propelled by four (quadrotor) or six (hexrotor) rotors and uses a control system to balance the thrust of each rotor to control drone lift and yaw, pitch, and roll angles. In this manner, the controller can produce stable flight of the drone (Patel *et al.,* 2013).

Figure 11. Hexrotor (AirRobot AR-200) Carrying a Vision Camera for Remote Detection of Weeds.

The use of multirotor drones for agriculture has increased drastically in recent years, and several specific designs have been developed (Gonzalez-De-Santos *et al.*, 2016) (see Figure 11). Equipped with the appropriate sensors (vision, infrared, multispectral, and hyperspectral cameras, *etc.*), drones allow farmers to obtain data (vegetation, leaf area, and reflectance indexes) from their fields to study delicate changes in crops that cannot be detected by scouting the ground. These data permit farmers to infer crop diseases, plagues, nutrient insufficiencies, water stress or excess, and other circumstances that could affect crop growth. With this information, farmers can plan possible remedies (irrigation, fertilization, weeds control, *etc.*). Drones will definitely be used intensively in intelligent farms and must be designed to include the new technologies to be used in intelligent factories.

e) Multirobot Systems and Fleets

Many research efforts have developed specifically designed autonomous applications for agriculture (Nørremark *et al.*, 2008, Bakker *et al.*, 2010 and Nagasaka *et al.*, 2009) and many other efforts are aimed at operating groups of vehicles under unified control. This approach embodies the concept of multirobot systems, which constitute a forward step in automating agricultural activities. Multirobot systems use several mobile robots to execute different tasks cooperatively and collaboratively. Using groups of robots, in which every robot interacts with others to achieve a well-defined objective, is an emerging and necessary concept for applying autonomous systems to daily agricultural tasks (Peleg, 2005, Blackmore *et al.*, 2005, Cheung *et al.*, 2008 and Sørensen and Bochtis, 2010).

The theoretical foundations of multirobot systems or robot fleets have been studied in recent years (Bautin *et al.*, 2020 and Bouraqadi *et al.*, 2020), but the first experimental fleets were tested only recently; no commercial equipment is available yet. The RHEA system was one of the first attempts to build a fleet of robots for agriculture (Gonzalez-De-Santos *et al.*, 2016) and focused on the design, development, and testing of a new generation of automatic and robotic systems for both chemical and physical effective weed management (see Figure 12). The fleet consisted of both aerial and ground vehicles equipped with a machine vision system to detect weeds. The ground robots were also equipped with agricultural implements for treatment applications. The RHEA system showed that multirobot systems enable the use of small vehicles that accomplish work volumes similar to that of a large machine but provide the following additional benefits:

1. More accurate positioning during operation;
2. An intrinsically lighter weight than conventional machines, which reduces soil compaction and improves vehicle safety for people, crops, and the robot itself–all of which are features currently requested for agricultural equipment (Blackmore *et al.*, 2001)
3. Easier acquisition because the multirobot approach allows farmers to acquire high-technology equipment in an incremental manner;

4. increased fault tolerance–a failure in a small robot means one less robot at work, while a failure in a large vehicle halts the entire field operation; and
5. Mission coordination and reconfiguration–at any time, the fleet behavior can be changed to optimize the mission or accommodate sudden changes in field conditions.

Figure 12. A Fleet of Robots (the RHEA fleet).

An important limitation of multirobot systems is that the total number of devices (*e.g.*, sensors, actuators, and computers/controllers) increases according to the number of robots; thus, a large number of robots can result in complex and expensive systems that are attractive only when applied to high-value crops (*e.g.*, grapes, peppers, and golf course turf).

Chapter 23

Human-Robot Interaction

The challenges for interaction range from domain independent aspects such as intuitive designs, immersive displays (*e.g.* Virtual and Augmented Reality) and tactile feedback, to very specific challenges stemming from the in-field conditions. Examples include the design of suitable interaction devices that are operable under harsh conditions, with constrained dexterity and precision of the operators, *e.g.* workers wearing gloves or having wet and muddy hands, or to guarantee the safety of often large and heavy semi-autonomous machinery in an environment shared with human workers. In contrast to robots in factories, where working areas can be fenced off when a robot is in operation, agricultural robots are limited by the absence of safety infrastructure in the fields, and require new innovative solutions.

1. Human-Robot Collaboration

Robots closely collaborating with humans so-called cobots (Akella *et al.*, 1999) are delivering real step changes in many industrial sectors, and are anticipated to be vital to automation in agriculture. Use cases range from farm in-field logistics (transportation), where efficient and safe hand-over of goods and produce needs to be facilitated, to applications enhancing animal and crop welfare by means of integrated monitoring and intervention delivery. An illustrative example is the RASberry project at the University of Lincoln, where human pickers of strawberries are supported by mobile robots acting as transporters. While some tasks for cobots require physical interaction between robots and humans, in other areas robots can act as a mediator or provide a remote presence for agronomists and farmers. Therefore a focus on intuitive and ergonomically appropriate interfaces and interaction design is needed. Concepts of shared autonomy and control, allowing operators to exercise control from remote locations over a potentially heterogeneous (ground, air, water) fleet of semi-autonomous robots, will also be important. As the technology matures, and in particular for safety-critical tasks, various levels of shared autonomy will be seen, where the human operator guides the high-level execution, while the robotic system performs the required sensorimotor coordination on the ground. The fan-out (Olsen and Wood, 2004), or number of robots a human can control simultaneously, will help drive the mixture of human supervisors and robot agents in such a paradigm.

Physical Human Robot Interaction (pHRI).

2. Safe Human-Robot Interaction

By relying on humans as supervisors, the autonomy levels, and associated risks and design complexities, can also be improved. Human supervision will be a vital safety factor for most agri-robotic systems for the foreseeable future, while the technology develops towards higher levels of autonomy. The robotic systems will also be learning and adapting to task and farm-specific constraints. Human and robot collaborators will therefore likely be mutually adapting to each other, in order to maximise performance. Approaches to safe physical Human Robot Interaction (pHRI) (Haddadin and Croft, 2016 and Cherubini *et al.,* 2016) include supervisory systems to monitor the interaction and adjust the behaviour of the robot if an unsafe situation is identified. This typically involves slowing, or completely stopping the robot, to prevent accidents. However, this approach can significantly reduce productivity as the robot is not working to its full potential. Current research aims to improve on this approach by allowing robots to identify and predict unsafe situations, and then to adapt and adjust their operation to continue the task in a manner that allows both productivity and safety to be maintained (Pereira and Althoff, 2017). A further approach to ensuring safe pHRI is to design robot systems which are inherently safe, meaning that if collisions occur between human and machine, injury will not result. The aim is to replicate the safe interaction that occurs when multiple people work collaboratively. This requires a change away from heavy, rigid and high inertia robots to systems which are more akin to biological creatures. Again this is a challenge that the new field of soft robotics may be able to address.

Chapter 24

Learning and Adaptation of Robots in Agriculture

Artificial intelligence technologies, especially in machine learning, are expected to play a major role in most of the robotic technology areas, and will be essential enablers for agricultural robots. Agricultural environments are subject to changes throughout the lifetime of a robotic system. For example, there may be new crop varieties, weeds, pests, diseases, treatments, legislation, climate change, *etc.*, as well as new implements and robotic technologies. In AI terms this means dealing with an open world, so techniques to enable adaptation during operation rather than at the design phase will be crucial. Techniques that allow robots to learn from experience include reinforcement learning, learning from demonstration, and transfer learning to exploit prior knowledge, *e.g.* from another domain or task. Ongoing research is investigating deep learning methods (Kamilaris and Prenafeta-Boldú, 2018), especially in perception-related tasks involving the interpretation of sensor data, including recognition and segmentation tasks in automated weeding and fruit picking. Robots will also need to leverage human knowledge, especially when facing situations that were not foreseen at design time. This additional input might be given by end-users, maintainers, and/or domain experts. It might also be provided through direct control (*i.e.* teleoperation), natural interaction (*e.g.* via language or gestures) or by the means of labelled examples and data sets. These developments will link naturally into the use of Big Data in smart farming (Wolfert *et al.*, 2017) alongside the use of satellite imaging, UAVs and ground robots for more localised and richer, multimodal data collection. These developments coupled with cloud-based storage will create an abundance of information that could potentially be utilised for smart planning and control of agriculture. An important requirement is the standardisation of data to ease the exchange between robots, domains, farms, countries and companies. RAS technologies have the potential to revolutionise all agricultural sectors. The nature of their contribution will differ across agricultural types, *i.e.* crops, livestock and aquaculture, and from phenotyping through to primary production activities. Across this spectrum there are many ways that robots could contribute, both economically (*e.g.* growing and harvesting more efficiently and cheaply), ecologically (*e.g.* reducing and eliminating the use of chemicals, while helping to maintain soil health) and ethically (*e.g.* increasing animal welfare via monitoring and timely intervention). In parallel, robotics may enable automation in the care of livestock and aquaculture or alternative growing systems, such as 'vertical' (protected-environment) farming or agroforestry systems combining agriculture and forestry on the same land. The technology challenge areas for agri-robotics tend to divide into two classes: (1) Breeding/Phenotyping (2) Farming/Primary Production. There is then a whole additional sector of post-harvest Agri-Food activities for robotics, which fall outside the remit of this white paper. Taking conventional terrestrial arable agriculture as a farming exemplar, the challenges can be illustrated as follows. Direct parallels may then be extrapolated for livestock and aquaculture as well as non-conventional farming systems, *e.g.* organic or 'vertical', however for the purposes of brevity these have been excluded from the narrative below.

Phenotyping

Laboratory

The identification and selection of genetics with beneficial abIoTic (*e.g.* drought or saline tolerance) or bIoTic (*e.g.* fungal viral or bacterial disease resistance) input traits and complementary output traits (*e.g.* nutrientto- biomass conversion, shelf-life, flavour or nutritional qualities) is conventionally undertaken within controlled laboratory environments. These breeding activities have seen a degree of robotic integration in recent years to reduce the reliance on manual intervention, but the cost and complexity of implementation, as well as the questionable reliability and technology-readiness of the current systems, has limited their uptake.

Field

While laboratory phenotypic screens may be important for identifying beneficial breeding lines to cross, they are only a proxy for the primary goal of determining how that crop may thrive in real-world conditions. Robotics opens up the potential for mass direct in-field phenotyping of crops under true farm conditions (Cseri *et al.,* 2013 and Burud *et al.,* 2017). Such uncontrolled 'non-laboratory' systems raise significant challenges over the singular identification of the specific trait that resulted in a beneficial phenotypic response. The robotic capability for repetitive and detailed assessment of the environment of individual plants opens up the potential for a paradigm shift in the development of agri-genetics.

Breeding Activities of Robot.

Establishment and Seeding

Ploughing is one of the most important primary cultivation processes, and involves the inversion or mixing of top soil to prepare a suitable seed bed. Currently modern agriculture uses a huge amount of energy in ploughing: it is estimated that 80 per cent -90 per cent of the energy in traditional cultivation is used to repair the damage done by large tractors. Small, smart, electric robots provide an alternative solution, by avoiding excessive compaction of the soil in the first place, and performing micro-tillage using on-board implements. Nutrients could also be better targeted to the local environment of individual seeds using precision approaches. Seed placement and mapping could be further automated to optimise the density and seeding pattern with respect to the requirements for air, light, nutrients and ground moisture of the individual crop plants. Robotics will also have an important role to play in managing the inputs to primary production, including both monitoring and interventions, particularly for soil (Fentanes *et al.,* 2018) and water (Hitz *et al.,* 2012).

Crop Care

One of the main operations in crop management is scouting to collect timely and accurate information. Autonomous robots carrying a range of sensors to assess crop health and status could thus assist in cost-effective data collection. Both aerial and ground-based platforms and their combination (Walter *et al.,* 2018) could be utilised. Fusing data collected by different devices or obtained from sources with a wide range of temporal and spatial resolutions and automatically interpreting data impose a number of interesting research challenges. Weed mapping involves recording the position and density (biomass) of different weeds species using machine vision. The resulting weed map can be further interpreted into a treatment map. Robotic weeding is an active area of current research, investigating alternative methods to kill, remove or retard unwanted plants without causing damaging to the crop. Intra-row weeding is more difficult than inter-row weeding, as it requires precise positioning of the crop plant.

Chapter 25

Benefits of Automation and Robotics in Agriculture

1. Precision agriculture, with its focus on efficiency and efficacy of agricultural inputs and the spatial and temporal management of agricultural systems
2. Specific agricultural production systems, including those related to field crops, cotton, orchards and vineyards, and animal housing and production
3. Automation relative to specific inputs in agricultural production systems, such as nutrition management and automation, automation of pesticide application systems, and automated irrigation management with soil and canopy sensing
4. Liability issues with regard to surrounding awareness and worksite management

Postharvest automation–perhaps the most advanced component of agricultural production in terms of automation and an important factor in global agriculture (Qin Zhang, 2013).

Automated agriculture is a diverse sphere as of now, so we are going to cover the latest innovations and the derived benefits.

1. Consumer Preferences

The overload of fast food and other unhealthy meals on the market has led to a demand for healthier food products. The impact that unhealthy food has had on our bodies has led to a "back to basics" approach: consumers all over the world are craving for plant-based wholesome food. Farmers should provide it in large amounts to meet the new consumer demand. Consumer preferences are shifting towards organic and sustainably-produced products. With automation technology, produce reaches consumers faster, fresher, and more sustainably. Increase in productivity from automation increases the yield and rate of production, therefore reducing costs for consumers.

2. Labor Shortage

You will hardly find a teenager nowadays who dreams of being a farmer. The truth is, the world population is seeking to live a more urban lifestyle, so the farming business has to deal with the labor shortage. Furthermore, modern agricultural practices have changed over the last decade, which makes it harder to train new workers.

3. Labor Efficiency

Labor is over 50 per cent of the cost to grow a farm and 55 per cent of farmers say they are impacted by labor shortages. Because of this, 31 per cent of farmers are moving to less labor-intensive crops. However, there is huge potential with harvest robots. Routine tasks can be automated with robotics technology, reducing labor costs and manpower needed amidst a labor shortage in the agriculture industry. A single strawberry robot harvester has the potential to pick a 25-acre area in 3 days and replace 30 farm workers.

Labour Engaged in Weeding.

4. Ecological Responsibility

The general public and local authorities expect agricultural companies to be eco-friendly and demonstrate a responsible approach to their work. Farmers should opt for a reduced usage of chemical components, in particular, pesticides. Business owners in the agricultural sector are looking for a solution in response to this social and ecological demand, and they can find it in farm automation.

1. One agrobot can change the future. Smart farm automation technology has great potential to be tremendously impactful. The solution to numerous global problems and the creation of eco-friendly approaches lies in this field of technology.
2. The agriculture industry meets consumers' demands.
3. We have mentioned that consumer demands and their preferences are changing. We all want food to be fresh and be quickly delivered to shops and markets. Agrobots allow farmers to save time and money.
4. The labor shortage in the industry is decreasing.
5. Labor in some farming branches can take up to 50 per cent of the overall cost. The labor shortage makes this situation even more dramatic. Using robots for seeding, harvesting, watering, and monitoring decreases the need to hire employees for numerous monotonous tasks.
6. Agriculture is becoming more eco-friendly.
7. The usage of pesticides frightens lots of people that are why agriculture is gaining an infamous reputation. Robots are shifting the rules of the game towards an environmentally-friendly approach: they can add pesticides with great precision, which is almost impossible when done by an actual human (*https://easternpeak.com* › blog › smart-farming-how-auto.)

5. Reduced Environmental Footprint

Farm automation practices can make agriculture more profitable while also reducing the ecological footprint of farming at the same time. Site-specific application software can reduce the amount of pesticides and fertilizer used while also reducing greenhouse gas emissions. By applying the automation and smart farming techniques we can change the action according to the crop, weather conditions, and soil, *etc.*

1. In the field of irrigation automation and smart systems can irrigate the horticulture land, parks, gardens, and golf courses efficiently and reduce the wastage of water.
2. The uses of automation and smart systems also help in food safety by the selection of appropriate chemicals.
3. The weed control system is also benefited by these technologies.
4. The application of multitasking robots is also useful to make it fast and maintain quality.
5. In the smart farming also helps to maintain the humidity, temperature and irrigation process by using the sensors and apps (Chidananda Kashyap *et al.,* 2019)

Besides making work a lot easier for the farmers, here are some ways robots can prove to be advantageous in agriculture:

1. ***Job Variation*** – Farming robots can perform various repetitive jobs like weeding, sowing, fertilizing, and spraying. Efficient navigation using multiple sensors and AI helps in performing nonlinear tasks faster and more efficiently.
2. ***Bring Automation*** – Robots can act as an automatic self-control system and function as efficient fertilizer or liquid sprays using state-of-the-art navigation algorithms and machine vision. Autonomous tractors will also become more commonplace with the aid of technological advancements in machine vision systems, IoT connectivity and LiDar.
3. ***Fewer Errors*** – Compared to humans, robots can be programmed to make fewer errors. Besides, robots can eliminate work that can be dangerous to humans.
4. ***Higher Efficiency*** – As the robots can discern features using 3D computer vision and GPS, they efficiently manage geographical obstacles like trees, ponds, and rocks.
5. ***Monitor Agricultural Health*** – Robotic cameras can identify pests and parasites, detect weeds, and enable the farming robots to spray the specific area with pesticides.
6. ***Good ROI*** – By enhancing production yields, agricultural robots increase the return on investment, thereby making farming a profitable venture.

All in all, robots have proved to be handy in the industrial sector as they speed up operations and save human labor. Hence, they have tremendous scope in an industry with the maximum human activity, *i.e.*, agriculture. Robots in agriculture can reduce the dependency on human labor to a significant extent. And they improve the overall efficiency levels considerably with holistic farming approaches by leveraging emerging technologies like AI and ML (Mina Mesbahi, 2021).

6. Economic and Societal Factors

Robotics and Autonomous Systems (RAS) are set to transform many global industries. These technologies will have greatest impact on large sectors of the economy with relatively low productivity such as Agri-Food (food production from the farm through to and including manufacturing to the retail shelf). The UK Agri-Food chain, from primary farming through to retail, generates over £108bn p.a., with 3.7m employees in a truly international industry yielding £20bn of exports in 2016 (National Farmers Union report, 2017). The global food chain cannot be taken for granted: it is under pressure from global population growth and needs to drive productivity, climate change, inescapable political impacts of migration (*e.g.* Brexit and potential US migration restrictions), population drift from rural to urban regions, and the demographics of an aging global population in advanced economies including China. In the UK the uncertainty associated with Brexit is already affecting migrant worker confidence and availability. These issues, manifest via different mechanisms (demographics, urban drift, *etc.*), are now impacting on many sectors of the global Agri-Food industry. In addition, jobs in the Agri-Food sector can be physically demanding, repetitive in nature, conducted in adverse environments and relatively unrewarding. Given these circumstances the global Agri-Food sector could be transformed by advanced RAS technologies.

The recent Made Smarter Review (Maier, 2017) considered that digital technologies, including RAS, deployed in food manufacturing alone could add £58 bn of GVA to the UK economy over the next 13 years. Robotic automation would also help to attract skilled workers and graduates to the sector. These opportunities have been further recognized by national government and are outlined in the Industrial Strategy White Paper (HM Government, 2017). The Secretary of State for BEIS announced in late February 2018 a £90m ISCF Wave 2 investment (Transforming food production: from farm to fork) to support innovation, including robotics and digital systems to drive innovation in the Agri-Food chain. This follows an earlier £160m investment that funded the UK Agri-Tech Catalyst program (now near completion).

7. Environmental Benefits

As well as delivering economic benefits, such as increasing productivity and reducing waste throughout the food supply chain, developing a new focus for RAS within Agri-Food will have significant societal and environmental benefits. For example, the food chain uses 18 per cent of UK energy consumption (Tassou *et al.*, 2014) while high mass farm machinery is causing unsustainable compaction damage to our soils; meanwhile, many species of British wildlife face risk of extinction due to modern farming practices (Hayhow, 2016), including widespread use of herbicides and pesticides, industrialisation of machinery, and reduction in hedgerows and drainage due to increasing field sizes.

a) Water

Agriculture uses 70 per cent of all global fresh water supplies, and yet 4bn people live in global regions with water scarcity (Mekonnen and Hoekstra, 2016). In the UK, weather can play an unpredictable role in agriculture with short periods of drought or flooding in many rural areas. This unpredictability makes it difficult to put in fixed systems of drainage or irrigation at a justifiable cost. The issues of water are complex: we need to find new means to drive water use efficiency within agricultural systems, while onward diffuse pollution to water bodies has serious negative environmental impacts. DEFRA estimates that the UK cost of diffuse pollution (nitrate, phosphorous, pesticide and sediment run-off, *etc.*) amounts to c. £311m per annum (Department for Environment, Food and Rural Affairs, 2016). Field robots are already being deployed to help farmers measure, map and optimise water and irrigation use. Likewise robots that use precision technologies to apply fertilisers and pesticides within agricultural systems will reduce environmental impacts.

b) Pesticides

Agricultural systems globally are now highly reliant on the large-scale application of synthetic pesticides to control weeds, insects and diseases. An improved understanding of chemical safety and environmental impacts has led to multiple product withdrawals, reducing the number of active ingredients available to farmers. Furthermore, ever higher regulation and registration costs have reduced the number of new pesticides entering the agricultural market. There is therefore a global need to find new ways to produce crops that do not require or reduce the use of pesticides. There are now a number of crop weeding robots that reduce the need for herbicides by deploying camera-guided hoes (Tillett *et al.*, 2008) precision sprayers (Binch and Fox, 2017) or lasers (Mathiassen *et al.*, 2006) to manage weeds. Although in its infancy, this technology shows great promise. In addition, novel sensors deployed on robots can reduce pesticide use by both detecting pests and diseases and precisely targeting the application of insecticides and fungicides. Robots could also be deployed as part of integrated pest management systems, for example, for the accurate and low-cost dispersal of biopesticides to counteract crop pests and diseases.

c) Electrification of Farm Vehicles and Implements

The proliferation of electric motors and actuators in applications ranging from industrial processes to modern passenger aircraft and cars is indicative of the drive away from mechanical traction and actuation systems to electrically based systems. Large diesel vehicles are likely to remain in practical use for many years to come, however the optimum use of such vehicles has been to go as large as possible, which in itself leads to issues such as soil compaction and "brute force" delivery of fertilisers, herbicides and pesticides. The migration from monolithic, fossil-fuel-based agricultural platforms to fleets of smaller electric powered robotic platforms offers the possibility of much lower emissions with locally generated power. Recent years have seen an increase in the use of agricultural land for solar photovoltaic, wind turbines and anaerobic digestion plants. Therefore, the potential for dual use of not only the land, but also the electricity generated, is of interest to the agricultural robotics community. Many agricultural implements are driven directly from the prime mover (often a tractor) via a mechanical linkage. By using electric drives the efficiency can be much higher and the whole system made safer as a result. One of the most common sources of injury and death on the farm is the mechanical linkages in large farm machinery. Therefore there are potentially major health and safety benefits to electrification and automation of farm equipment.

8. Livestock and Aquaculture

Robotics and autonomous systems for livestock at the farm level, robotic systems are now commonly deployed for milking animals (International Federation of Robotics, 2017). The take-up is a relatively small percentage at the moment, but an EU foresight study predicts that around 50 per cent of all European herds will be milked by robots by 2025(European Parliamentary Research Service, 2016). Robotic systems are starting to perform tasks around the farm, such as removing waste from animal cubicle pens, carrying and moving feedstuffs, *etc.* Systems are in use and under development for autonomously monitoring livestock and collecting field data, all commercially useful for efficient and productive livestock farming. There are further opportunities to apply more advanced sensor technologies, combined with more autonomous systems, to perform tasks on the farm. This applies to both extensive production and intensive (indoor) systems. Extensive livestock utilise 45 per cent of UK's area that is grassland and not fit for crop production, and derive food products from this resource. Thus management of this feed resource is also important. A further application for robotic systems concerns the management of farmed animals, such as dairy cattle, pigs and chickens, where intervention via the provision of appropriate and timely data can help reduce waste and environmental pollution as well as improve animal welfare and productivity on the farm. Welfare accreditation schemes and measures undertaken for assurance purposes (RSPCA Assured, Red Tractor, *etc.*) rely heavily on systems-based approaches by inspectors observing the farm intermittently at a group level. Precision farming on the other hand has the potential to offer animal-centric health and welfare assessment that would operate continuously to assess the condition and state of individual animals. Other wider societal benefits may follow in terms of improved working conditions, a stronger more competitive UK agricultural sector (via improve feed conversion efficiency per kg of meat produced) and better food security, providing consumers with greater access to lower-cost, higher quality meat products. Farmers must constantly monitor their animals and their setting in order to ensure animal health and maintain a comfortable, stress-free environment for optimal production. Though the UK has some of the highest animal-welfare standards in the world, there are pressing concerns over maintaining this post-Brexit due to potentially cheaper low welfare imports from outside Europe. Other constraints (*e.g.* high feed costs, environmental regulation, consumer concern over animal welfare and food security, long-term publichealth concerns such as antimicrobial resistance) add to the pressure in this cost competitive sector, driving governments and consumers to demand greater supply-chain control. Unfortunately human monitoring involves many limitations, including contamination and farm-worker health risks, and provision of only limited frequency, resolution and fidelity of data. It is also a slow, costly and labour-intensive process. Automation offers the potential for continuous data capture, allowing timelier and effective intervention, improved animal welfare and reduced production costs. Robotics and autonomous systems for Aquaculture production is already a vertically integrated and professional supply chain, but operates in an environment with a number of challenges that limit production, where sensors and robotic systems can play a role. Any systems deployed are naturally required to be more robust to extreme conditions and environments. The environment for aquaculture is often hostile and difficult to access by human operators with remote locations and inclement weather, with access only by small boat, leading to high operating costs and significant health and safety issues. The use of autonomous sensing and remote operation could significantly reduce the requirement for an on-site human presence making such facilities safer and easier to manage. Major challenges include environmental and health issues, such as algal blooms, sea lice and gill diseases. Could robots be used to monitor and offer a mode of treatment to control parasites, for example? Could robots assist in benign control of seals and other protected species? Are there autonomous means of maintenance for the required infrastructure, especially as the industry has started to move recently to "higher energy" sites? Could more precise understanding of the environment and behaviour of fish lead to better management control and therefore better productivity? These and many other challenges exist in the aquaculture sector, even those systems that are land-based.

9. Non-Conventional Closed ('Vertical') Farming

'Vertical farming' systems utilise indoor farming techniques and closed environments where all environmental factors, including nutrients, temperature, humidity and lighting, can be controlled (Kozai *et al.,* 2015). The co-design of robotics with specialized sensing and crop genetics, where plants have been bred to take advantage of the closed environments, is now enabling these systems to move from niche low-volume markets, such as localised production of herbs for high end restaurants, to mainstream reliable delivery of volume produce at 'Amazon' scales. Such large-area enclosed farming systems have been referred to previously as 'vertical farms' to emphasise the high density configuration of plants. In reality, robotic sensing and effector units, typically gantry mounted, may be engineered to accommodate plants in any combination of vertical or horizontal modules to optimize the use of space. Thus, such systems are particularly well suited for high-density urban environments, where the production of crops with guaranteed growth times near the point-of-purchase enables minimisation of both waste, due to over-production, and the carbon footprint associated with long-distance transport chains. Rural economies may also gain benefits from access to locally derive and nutritionally advantageous fresh produce lines that have traditionally been cost-prohibitive or logistically impossible to deliver. The required components for 'vertical farms' include temperature and humidity control, balanced crop-nutrient chemistry, process engineering of hydroponics or other growing media, semiconductor illumination, non-invasive sensing and robotics. Now these engineering elements are all realisable, the final ingredient required is the design of crops specifically for such systems, as opposed to their translation from field or polytunnel varieties. Within these controlled environments, crops can be bred specifically to maximise yield or other desired output traits, without the need to breed for other factors, *e.g.* resilience to specific pests, weeds, *etc.*, as in conventional growing systems. Robotic systems would then allow novel output traits to be nurtured and bred into future produce, such as beneficial health and nutritional aspects, or to minimise the energy and waste in any downstream processing. A further benefit would be to deliver plants that are more compatible with autonomous plant care and harvesting. While 'vertical' production systems may not be economically or environmentally competitive with the existing systems for the foreseeable future, the potential for such units within urban and industrial environments cannot be considered in isolation. As plants are potential sinks for effluents, heat and excess energy production then the rationale for such food production facilities needs to be considered alongside their potentially beneficial effects that would otherwise have to be dealt with via alternate, less sustainable means. An exemplar is the positioning of British Sugar Ltd. as the biggest UK tomato producer as a consequence of being a sink to the carbon dioxide release from the sugar processing and the exothermic reactions which promote tomato growth. By appropriate design of the automation and robotics within non-conventional vertical farming units, the economic arguments for installing such systems may be overlaid with their ability to solve issues within parallel sectors such as exploiting alternative waste and excess energy streams.

10. Food Manufacturing and Processing

While post-harvest activities are beyond the main focus of this white paper, we note that the need or new research and innovations across Agri-Food do not stop at the farm gate. For example, the meat sector presents particular challenges in productivity, due to the increasing difficulties in finding skilled workers with the qualifications needed to carry out meat cutting tasks. Robotisation of these jobs has thus become an important objective for companies looking to improve the safety and health of workers, as well as finding new solutions to mitigate the increasing production costs linked to current and future labour shortages. Collaborative robotics (cobots), where robots work together with humans, presents an alternative solution to help increase productivity, improve health and safety, and attract skilled workers and graduates to the food processing industry. Agricultural robotics could also facilitate earlier labelling and tracking of food products throughout the manufacturing supply chain, bringing numerous benefits such as improved information for consumers on where their food is from and faster action to mitigate food safety issues. In turn, the food metadata could be fed back to the field operations to further improve the primary production. In general, there may be many potential synergies between agricultural robotics and the downstream processing of agricultural products in the food chain, where whole supply chain efficiencies could be unlocked through future application of RAS technologies. Common challenges that span the whole Agri-Food supply chain include the demand for soft robotics; human-robot collaboration; safety of both food products, people and other assets; automation of intra-logistics, from in-field transportation to packhouse to warehouse operations; sensing and image interpretation for analysis and manipulation of complex food products; and long-term autonomy, requiring the development of robust, fault-tolerant systems able to operate 24/7 in challenging field and factory environments.

11. Technological Focus

The recent focus of the agri-robotics community has been to identify applications where the automation of repetitive tasks is more efficient or effective than a traditional human or large machine approach (EU Robotics AISBL 2014 and Bechar and Vigneault, 2016). Research is needed into robotic platforms that can operate close to the crop (either on the ground or at elevation) and advanced manipulation, especially with interactive or tactile properties, *e.g.* for picking soft fruit. The use of heterogeneous "multi-modal" platforms that combine ground-based and aerial vehicles provides opportunities for targeted support and intelligence for the individual platforms, plus the ability for human operators to have an "eye in the sky" for observation and mission planning. Collaborative and cooperative behavior becomes advantageous for large-scale arable and fruit crops as tasks can be performed in parallel, giving economies of scale. Land management is a specific issue of concern in the UK landscape, given the issues of fertilisation, water management and carbon content in the soil, so the use of advanced sensing and soil management was using remote platforms including robotics will be increasingly important. Additionally, the use of robotics for livestock management is a specific opportunity for the deployment of autonomous platforms, as has already begun in automated milking stations, and with potential applications for raising animals in fields, barns, sheds and aquaculture, or fish farms.

Picking Soft Fruits.

12. Related Areas

There is a plethora of related areas that are already using automation (such as in large parts of the industrial food production in the UK), and research is needed to investigate how this can be more tightly integrated into the agriculture industry. The food chain is managed using complex food production and software systems that rely on accurate data about all aspects of the location, quality and quantity of agricultural foodstuffs. Robotics and automation are already being used extensively in the processing side of the food industry; however this is not being leveraged to the same extent on the production side in the field. The application of large data sets in combination with remote sensing (as used in tracking raw materials for industrial production, *e.g.* with RFID tags) to optimise the quantity and quality of crops or livestock produced has the potential to revolutionise the UK agricultural sector. Technologies from related areas including the Internet of Things, Big Data and artificial intelligence can be used alongside autonomous systems technologies to automatically fuse and interpret collected data, assess crop status, and automatically plan effective and timely interventions in response to sudden events and the change of crop conditions (*e.g.* weather, diseases and pests).

13. Technology Vision

Our long-term technology vision encompasses a new generation of smart, flexible, robust, compliant, interconnected robotic systems working seamlessly alongside their human co-workers in farms and food factories. Teams of multi-model, interoperable robotic systems will self-organise and coordinate their activities alongside and within existing Agri-Food systems. Electric farm and factory robots with interchangeable tools, including low-tillage solutions, novel soft robotic grasping technologies and sensors, will support the sustainable intensification of agriculture and drive manufacturing productivity throughout the food chain. Future agri-robotic systems will deploy artificial intelligence and machine learning techniques to increase their own productivity. Meanwhile, investigation of alternative systems for food production, including innovations from areas such as vertical farming, will further help to address the sustainable intensification of agriculture, while protecting the environment, food quality and health. A vital aspect of making this transition effective is the clear demonstration of economic benefits, which has always been the primary driver of change to the agricultural community. Facilitating the transition to automation While full automation is often hailed as the ultimate aim in technological development, and the future agriculture systems may look very different from those of today, only very few large companies can afford the disruption of full automation. So to achieve this long-term vision will require a gradual transition from the current farming practices, and most farmers will need technologies than can be introduced step by step, alongside and within their existing systems. Furthermore, while some emerging robotic technologies are already achieving or approaching the robustness and cost-effectiveness required for real-world deployment, other technologies are not yet at that stage. For example, soft fruit picking still requires fundamental research in sensing, manipulation and soft robotics. Thus, at least in the short term, the collaboration of humans and robots is fundamental to increased productivity and food quality (Agricultural robots for field operations, part 2, 2016). Thus, much of the research needed in the short- and medium-term should focus on facilitating the "transition to automation", with mixed systems likely to dominate in the coming years, benefitting from collaboration between humans and robots, with combinations of electric, diesel and hybrid powered vehicles, as the required technologies for electrification mature and become ready for market. In the short-term, progress may depend on retrofitting of existing farm vehicles. For example, a human-driven tractor could tow a variety of robotic implements for different field operations such as selective harvesting or weeding. In the longer term, autonomous robotic vehicles will start to replace the legacy vehicles. This trajectory will also enable the UK vehicle and implement manufacturers to develop new products that span the transition from the current diesel powered farm vehicles to the robotic farming systems of the future. The UK is well placed to implement these changes due to its strong automotive sector in industrial and agricultural vehicles, with extensive infrastructure already in place. Small, smart, interconnected, light machines one advantage of modern robotics is their ability to be built using low-cost, lightweight and smart components. Due to their prevalence in consumer electronics, such as mobile phones, gaming consoles and mobile computing (laptops, tablets *etc.*), high quality cameras and embedded processors can be built in to many platforms at very low cost. New materials and fabrication techniques such as additive manufacturing and advanced composites are also making the manufacture and deployment of robotic platforms much cheaper and decoupled from a mainstream manufacturing process or supply chain. For example, a 3D printer located on a farm could be used to manufacture spare parts on demand, at very low cost, or even to improve the platform by adapting to the local conditions. Using collaborative and cooperative behaviour in a fleet of robots further provides the opportunity to spread tasks over multiple platforms and thereby reduce the damage caused by heavy conventional agricultural platforms on the soil or existing crops. The robotic fleet can also take advantage of multiple data sources to calibrate the task, reduce waste and focus on areas of greatest need, potentially reducing fertiliser costs and environmental impact.

Chapter 26

Advantages of Agricultural Robots

- The robots do not get sick or tired and they do not need the time off. They can operate with closer tolerances (so, every round is at full field capacity). They offer fewer errors and at higher speeds and the higher quality products can be sensed by the machines accurately.
- As the machines can be made lighter and cheaper if the drives seat, controls and cab can be eliminated. The robots can be used in various fields in the agriculture. The robots can easily work around the tree, the rocks, the ponds and the other obstacles.
- The robots can reduce up to 80 per cent of farms use of pesticide. The robots may perform more or different tasks in the future. The robots can create the jobs for the peoples who have to make the robots and who have to fix the robots.
- The robots have many fields of application in the agriculture such as the Merlin Robot Miller, Rosphere, Harvest Automation, Orange Harvester, lettuce bot and the weeder.
- Another field of application is the horticulture. One horticultural application is the development of RV 100 by Harvest Automation Inc. RV 100 is used in handling and organizing the potted plants include the spacing capabilities, the collection and the consolidation, RV 100 for this task offers high placement accuracy, the autonomous outdoor and the indoor functions, and reduced production costs.
- The idea of applying the robotics technology in the agriculture in very new. In the agriculture, the opportunities for the robot-enhanced productivity are immense and the robots are appearing on the farms in various guises and in increasing numbers.

- You can expect the robots performing the agricultural operations autonomously such as the spraying and the mechanical weed control, the fruit picking, allowing the farmers to reduce the environmental impact, increase the precision and efficiency and manage individual plants in novel ways.
- The robotics are spreading every day to cover further domains, as the chance of replacing the human operators provides the effective solutions with return on the investment. The robots do the difficult tasks that are harmful to the health of the workers so. They offer the safety for the workers.
- The robots can protect the human workers from the harmful effects of handing the chemicals by the hand and through the system of high spraying and they can reduce up to 80 per cent of a farms use of pesticides.
- The fruits picking robots, the driver-less tractor/the sprayer and the sheep shearing robots are designed to replace the human labor, a lot of factors have to be considered such as the size and colour of the fruit to be picked, before the commencement of the task.
- The robots can be used for the other horticultural tasks like the pruning, the weeding, the spraying and the monitoring. They can be used in the livestock applications (the livestock robotics) such as the automatic milking, washing and castrating.
- The robots in the agricultural industry offer higher quality of fresh produce, lower production costs, and the smaller need for the manual labor. They can be used to automate the manual tasks such as the weed or the bracken spraying where the use of tractors and the other manned vehicles is too dangerous for the operators.
- The rice planters plant rice in the rice fields. The robotic milker milk cows in the use barns or in the milking factories and the fruit pickers pick the fruits in the fruit field and the robots are programmed depending on their tasks. The robotic milker and the fruit picker are multi-functional while the rice planter is not.
- The heavy chemi also or the drugs dispensers, manure or the fertilizers spreaders are the activities that concerned by the deployment of unmanned options. The agricultural robots use the combination of the advanced sensors, the cameras, the software and the technology.

Chapter 27

Agricultural Robots Disadvantages

- It costs a lot of money to make or buy the robots. They need maintenance to keep them running
- The farm workers can lose their jobs and no alternative for their survival due to machanization of farm operations
- The robots can change the culture/the emotional appeal of the agriculture and the energy issues are costly.
- The robots may be more acceptable to the non-farm community. The jobs in the agriculture require intelligence and quick, and the robots can be substituted with the human operator (Heba Soffar, 2016).
- Just as farming robots are advantageous; they come with distinct disadvantages, as well. Robots are specialized machinery. Hence, they are costly to acquire and, being machines, require a great deal of care to keep them functioning.
- The Agri-robots need a great deal of research and development to be continuously improved and updated. It involves massive R and D investments and also not easily accessible for the small and marginal farmers.
- On the other hand, conventionally, farming has been a manual activity over the years, and getting used to agricultural robots will be a cultural and emotional change.
- Automation and robotics only feasible for big farmers and landlords were not for small and marginal farmers.

- Automation and robotics only applicable for the developed countries and not practically feasible for developing and high populated countries like India.
- For under developed counties no chance for automation and robotics only depend on labours for their day to day agricultural operations.
- Maintanance cost and effective handling of robots are all challenging among the farming communities.

Chapter 28

Challenges of Farm Automation and Robotics

Still, there are challenges of farm automation that must be overcome. High costs to adopt robotic technologies present a huge barrier to entry for farmers, especially in developing countries. For example, robotic planters must carry water or pesticides with significant weight; the hardware must be built differently, which results in higher costs to make it a larger size. Technical issues and breakdown of equipment also present high costs to fix for such specialized equipment. In order to fully utilize farm automation, farmers will need to combine their knowledge and experience with these new technologies.

Challenges of Farm Automation.

We are only at the early stages of farm automation technology, but it will be able to transform agriculture. It offers a path towards sustainable and more efficient agriculture by advancements of technologies, production systems, and software. Every year, automation technology becomes more sophisticated, and what was cutting-edge just a few years ago will become commonplace and cost-effective soon. The human element will always be a fundamental aspect of managing a farm, but fully autonomous vehicles and farm equipment are coming.

Weed control include vision-guided mechanical weeding, selective (micro-) spraying and laser weeding. Irrigation is another area where robots could assist in targeting water in the right place at the right time. Pre-harvest assessment and yield forecasting (Aggelopoulou *et al.*, 2010) by robotic sensory systems will further assist in choosing the right time to harvest the crop.

Selective Harvesting

Selective harvesting involves harvesting only those parts of the crop that meet certain quality or quantity thresholds (Bac *et al.*, 2014). Two criteria are needed: the ability to sense the required quality factor before harvest (in-field grading) and the ability to harvest the product of interest without damaging the remaining crop. Selective harvesting presents several challenges for current robotic technology, perhaps the foremost of which is how to perform autonomous sensorimotor coordination with noisy and incomplete sensory data in the complex agricultural environment. This will likely require enhanced machine vision for recognition, segmentation, spatial localisation and tracking, as well as specialist robotic technology that is both robust and precise. Precision has traditionally been provided through stiff and easy-to-model robot mechanics. However, the increased computational resources available on robot platforms could enable some of the burden of precision to be handled through software and sensing, while allowing the robotic harvesting implements to become more passively compliant, safe and robust. Another challenge is to determine how much the robotic system should be adapted to a given crop and growing environment, and how much the growing environment should be adapted to enable better selective harvesting with robotic systems. There are interesting trade-offs to be made in this space, and which might differ from crop to crop. A related question is how to make sure the utilisation of expensive robotic hardware is maximised through the year, in particular for seasonal crops like soft fruits.

Finally, there may be many additional opportunities for serendipitous parallel usage of autonomous robots alongside other field operations, for example, helping to monitor and secure farm equipment from theft and criminal damage, or protecting the sensitive habitats and species of wildlife that often coexist with agriculture.

Chapter 29

Future Scope for Automation, Robotics and Wireless System Networks in Agriculture

- The farmers who are young will make more investments in automation with much interest than the elder farmers.
- The technology which is new has to be introduced slowly with time. Slowly the agriculture sector is moving towards precision farming in which management will we done on the basis of individual plant.
- Deep learning and other extend methods are used to detect the plant or flower type, this will help farmers to provide favourable environment to the plant for sustainable growth. Eventually the production of more customised fruits and plants will grow, which leads to an increase in the diversity of products and production method.
- Artificial intelligence techniques are growing at a rapid scale and it can be used to detect disease of plants or any unwanted weed in the farm by using CNN, RNN or any other computational network.
- Green house farming can provide a particular environment to the plants but it is not possible without human intervention.
- Here, wireless technology and IOT comes in the run and using the latest communication protocols and sensors we can implement weather monitoring and control without human presence in the farm.

- Harvesting of fruits and crops can also be incorporated by robots which are specialized in working round the clock for quick harvesting.
- Application of robotics are vast in farming such as the robots can be used in seeding and planting, fertilizing and irrigation, crop weeding and spraying, harvesting and shepherding. To complete the same work in many cases, it would take approximately 25 to 30 workers.
- Thermal Imaging can also be implemented by using drones and thermal camera in it. The drones monitor the farm and gives continuous real time data of the field so that the farmers could know in which area of the field the water quantity is less and can only start irrigation in that particular area. This will prevent water flooding or scarcity of water in the field and the crops get advent amount of water all the time. Many different integrated approaches can be used to provide a viable environment and increased growth.
- Agriculture monitoring is the much necessitate reducing human interventions in practice. Day by day demand for food is reaching its high peak and the without execution of the modern methods in agriculture it is very hard to achieve the increasing demand.
- Agriculture monitoring is the prime concern as it helps to reduce labour and increase the production.
- Artificial Intelligence has been implemented in crop selection and to help the farmer in the selection of the fertilizers.
- With the help of the database which the user has gathered and specified to the system, the machine communicates among themselve to decide which crop is suitable for harvesting and also the fertilizers which promote the maximum growth. Deep learning has wide reach and its application.
- The technology featured above might give one the impression we are on the verge of a fully automated agricultural system. However, farmers surveyed by Case IH indicated that while they are optimistic about machines tackling limited chores, they are less enthused by the prospect of robots planning and planting crops–tasks that set the stage for the entire season. Mark Young, CTO of The Climate Corporation, told *Successful Farming* that he thinks it is unlikely that robots will replace farmers anytime soon, though their promise of efficiency and easing resource constraints suggest that they most certainly have a place in the field–pun intended.
- Another expert, Yael Edan, a professor at Agricultural, Biological and Cognitive (ABC) Robotics at Ben-Gurion University in Israel, concurs.
- "We don't expect these robots to completely replace humans in the fields," Edan told *Digital Trends*. "We expect them to help with tasks humans can't and don't want to do." Nonetheless, Edan estimates we will see broader

adoption of agro-bots in commercial farming over the next five years (Aswath, 2017).

- Agricultural robots or agribots are changing the look, feel and pace of traditional farming practices. Crop harvesting is poised to significantly impact the agricultural sector over the next decade. Essentially, where consumer demand and labour requirements are the greatest, automation will prove most useful. For example, roughly 320,000 acres of apples are grown in the U.S. annually and each acre requires between 250 and 350 man-hours per acre.
- However, it is important to keep in mind that there will be a learning curve as these technologies improve in their sensitivity and operation capacity. The industry appears to be inching towards large-scale efforts, so robot developers will need to keep this growing trend in mind.
- Drone technology has reportedly demonstrated some promising results as in the case of Agribotix. In comparison to other robotic categories where ROI is scarce, the company's openness to publish hard figures may reflect the sustainability of the this technology (we regularly publish reputable case studies of successful AI implementations in business, and we've found such published case studies to be relatively rare in the nascent world of agricultural robots).
- Advances in drone research will be critical for maintaining safety considerations. Agriculture authorities and investors alike will require proof effectiveness and safety. We can anticipate more drone applications to attempt to come to market in the coming decade.
- Barring adverse weather, drones seem to have the easiest chance of becoming a commonplace agricultural application because – unlike their grounded counterparts – they don't need to navigate the unpredictable and messy world of sold and product.
- Grounded machines require dexterity in interacting with plants safely, while drones can survey a domain and record image data without requiring the same kind of significant progress needed for other robotics applications (case in point: the dexterity and discernment needed to pick strawberries at human speed is a much harder problem than simply flying over crops). We suspect that drones will be among the first widely adopted agricultural robots in the near term.
- While fully autonomous robots are a goal of many players in the agribot space, this should not overshadow the important role that humans will play in quality management. The evolving roles of robots and human laborers may potentially usher in new job roles that are more technical in nature.

- AI agribots are poised to present new opportunities for the agricultural sector and we anticipate increased implementation across multiple areas of the industry (Kumba Sennaar, 2019).
- Robots are intelligent automation systems to perform highly complex and unstructured work in agriculture. There are various subsystems which need to perform the intended task. Robotic systems for crop production need to be developed to operate under unstructured agricultural environments without reducing productivity and work quality. In that area, there has been considerable progress in the few last decades.
- The robotic system has been developed mainly for seeding, planting, spraying, grafting and harvesting. All the research has been carried in controlled or green house conditions. It was observed that the robotic system failed were there sudden change in working environment. Few attempts have been in India in the field of robotic application of agriculture, but they all are developed at laboratory scale. With advancement of embedded system, computing, machine learning and artificial intelligence, attempt can be made to promote precision agriculture.

References

ABB. YuMi, 2018. Creating an Automated Future Together. Available online: https://new.abb.com/products/robotics/industrial-robots/yumi (accessed on 2 July 2020).

Abdul Montaqimt, 2018. Top 30 interesting farming automation technologies and companies, JUNE 28, 2018. https://roboticsandautomationnews.com › 2018/06/28.

Abeach, L. A. A, S. Nefti-Meziani, and S. Davis, 2017. "Design of a variable stiffness soft dexterous gripper," Soft Robotics, Jun 2017. [Online]. Available: https://doi.org/10.1089/soro.2016.0044.

Adamides G, Christou G, Katsanos C, Kostaras N, Xenos M, Hadzilacos T and Edan, Y. 2014. A Reality-Based Interaction Interface for An Agricultural Tele-operated Robot Sprayer. Agriculture, 65(2), 247-257.

Aggelopoulou, A. D, D. Bochtis, S. Fountas, K. C. Swain, T. A. Gemtos, and G. D. Nanos, 2010. "Yield prediction in apple orchards based on image processing," Precision Agriculture, vol. 12, no. 3, pp. 448–456, Aug 2010. [Online]. Available: https://doi. org/10.1007/s11119-010-9187.

AGREENCULTURE. 2020. Centéol and Movies. Available online: https://www.agreenculture.net/copy-ofchallenge- centeol (accessed on 2 July2020).

Agricultural robots for field operations. part 2, 2016. "Operations and systems," Biosystems Engineering, vol. 153, pp. 110–128, Jan 2017. [Online]. Available: https://doi. org/10.1016/j.biosystemseng.2016.11.004.

Akella, P, M. Peshkin, E. Colgate, W. Wannasuphoprasit, N. Nagesh, J. Wells, S. Holland, Pearson, and B. Peacock, 1999. "Cobots for the automobile assembly line," in

Proceedings 1999 IEEE International Conference on Robotics and Automation (Cat. No.99CH36288C). IEEE. [Online]. Available: https://doi. org/10.1109/robot.1999.770061.

Albani, D,.J. IJsselmuiden, R. Haken, and V. Trianni, 2017. "Monitoring and mapping with robot swarms for agricultural applications," 14th IEEE International Conference on Advanced Video and Signal Based Surveillance (AVSS). IEEE, Aug 2017. [Online]. Available: https://doi.org/10.1109/avss.2017.8078478.

Alex Owen-Hill, 2017. Top 10 Robotic Applications in the Agricultural Industry. Last updated on Aug 01,2017. Posted on Aug 01, 2017 7:00 AM.

Anupam Anand and Saravanan raj. 2019. Agritech Startups: The Ray of Hope in Indian Agriculture Discussion Paper 10. MANAGE-Centre for Agricultural Extension Innovations, Reforms, and Agripreneurship (CAEIRA).

Aswath, 2017. Sensors, Automation and Engineering; The Future of Farming In India.

Autopilot. 2020. Available online: http://www.trimble.com/Agriculture/autopilot. aspx (accessed on 2 July 2020).

AutoTrac. 2020. Available online: https://www.deere.com/en_INT/products/ (accessed on 2 July 2020).

Bac, C. W, E. J. Van Henten, J. Hemming, and Y. Edan, 2014. "Harvesting robots for high-value crops: State-of-the-art review and challenges ahead," Journal of Field Robotics, vol. 31, no. 6, pp. 888–911, Jul 2014. [Online]. Available: https://doi. org/10.1002/rob.21525.

Baerdemaeker, J. De. 2013. "Precision agriculture technology and robotics for good agricultural practices," IFAC Proc. Elsevier.

Bakker, T.; Van Asselt, K.; Bontsema, J.; Müller, J and Van Straten, G. 2010. A path following algorithm for mobile robots. Auton. Robot. 29, 85–97. [CrossRef].

Bargoti, S., Underwood, J., 2017. Deep Fruit Detection in Orchards. 2017 IEEE International Conference on Robotics and Automation (ICRA) 3626–3633.

Bargoti, S and Underwood, J.P. 2017. Image Segmentation for Fruit Detection and Yield Estimation in Apple Orchards. J. Field Robot., 34, 1039–1060.

Barnes, M, T. Duckett, G. Cielniak, G. Stroud, and G. Harper, 2010. "Visual detection of blemishes in potatoes using minimalist boosted classifiers," Journal of Food Engineering, vol. 98, no. 3, pp. 339–346, Jun 2010. [Online]. Available: https://doi.org/10.1016/j.jfoodeng.2010.01.010.

Bautin, A.; Simonin, O and Charpillet, F. 2020. Towards a communication free coordination for multi-robot exploration. In Proceedings of the 6th National Conference on Control Architectures of Robots, Grenoble, France, 24–25 May 2011; Available online: https://hal.inria.fr/inria-00599605/document (accessed on 2 July 2020).

Bawden, O.; Kulk, J.; Russell, R.; McCool, C.; English, A.; Dayoub, F.; Lehnert, C and Perez, T, 2017. Robot for weed species plant-specific management. J. Field Robot. 34, 1179–1199. [CrossRef]

Bechar, A and C. Vigneault, 2016. "Agricultural robots for field operations: Concepts and components," Biosystems Engineering, vol. 149, pp. 94–111, Sep 2016. [Online]. Available: https://doi.org/10.1016/j.biosystemseng.2016.06.014.

Bergerman, M.; Maeta, S.M.; Zhang, J.; Freitas, G.M.; Hamner, B.; Singh, S and Kantor, G. 2015. Robot Farmers: Autonomous Orchard Vehicles Help Tree Fruit Production. IEEE Robot. Autom. Mag. 22, 54–63. [CrossRef]

Bikram Jyoti, N S Chandel and K N Agarwal, 2020. Application of Robotics in Agriculture: An Indian Perspective, 8th Asian-Australian Conference on Precision Agriculture, Oct 14-17, 2019,

Binch, A and C. Fox, 2017. "Controlled comparison of machine vision algorithms for Rumex and Urtica detection in grassland," Computers and Electronics in Agriculture, vol. 140, pp. 123–138.

Blackmore, B.S.; Have, H and Fountas, S. 2001. A specification of behavioural requirements for an autonomous tractor. In Proceedings of the 6th International Symposium on Fruit, Nut and Vegetable Production Engineering Conference, Potsdam, Germany, 11–14 September 2001; pp. 25–36.

Blackmore, S.; Stout, B.;Wang, M and Runov, B. 2005. Robotic agriculture–the future of agricultural mechanisation. In Proceedings of the 5th European Conference on Precision Agriculture, Uppsala, Sweden, 9–12 June 2005; Sta_ord, J.V., Ed.;Wageningen Academic Publishers: Wageningen, The Netherlands, pp. 621–628.

Bogue, R. 2016. Robots poised to revolutionise agriculture. Ind. Robot. Int. J. 43, 450–456. [CrossRef].

Bosilj, P, T. Duckett, and G. Cielniak, 2018. "Connected attribute morphology for unified vegetation segmentation and classification in precision agriculture," Computers in Industry, Special Issue on Machine Vision for Outdoor Environments, vol. 98, pp. 226–240.

Bouraqadi, N.; Fabresse and L. Doniec, 2020. On fleet size optimization for multi-robot frontier-based exploration. In Proceedings of the 7th National Conference on Control Architectures of Robots, Nancy, France, 10–11 May 2012; pp. 1–9. Available online: http://car2012.loria.fr/files/2012/Bouraqadi-CAR2012.pdf (accessed on 2 July 2020). Agronomy 10, 1638 24 of 24.

Burud, I, G. Lange, M. Lillemo, E. Bleken, L. Grimstad, and P. J. From, 2017. "Exploring robots and UAVs as phenotyping tools in plant breeding," IFAC-PapersOnLine, vol. 50, no. 1, pp. 11 479– 11 484, Jul 2017. [Online]. Available: https://doi.org/10.1016/j. ifacol.2017.08.1591.

CASE. 2017. Available online: https://www.therobotreport.com/case-ih-displays-new-cab-less-concepttractor/(accessed on 2 July 2020).

Ceres, R.; Pons, J.; Jiménez, A.; Martín, J and Calderón, L. 1998. Design and implementation of an aided fruit-harvesting robot (Agribot). Ind. Robot Int. J. 25, 337–346.

Chamen, W. T., A. P. Moxey, W. Towers, B. Balana, and P. D. Hallett, 2015. "Mitigating arable soil compaction: A review and analysis of available cost and benefit data," Soil and Tillage Research, vol. 146, pp. 10–25, Mar. 2015 [Online]. Available: https://doi. org/10.1016/j.still.2014.09.011.

Cherubini, A, R. Passama, A. Crosnier, A. Lasnier, and P. Fraisse, 2016. "Collaborative manufacturing with physical human–robot interaction," Robotics and Computer-Integrated Manufacturing, vol. 40, pp. 1–13, Aug 2016. [Online]. Available: https://doi.org/10.1016/j. rcim.2015.12.007.

Cheung, B.K.-S.; Choy, K.; Li, C.-L.; Shi,W and Tang, J. 2008. Dynamic routing model and solution methods for fleet management with mobile technologies. Int. J. Prod. Econ. 113, 694–705. [CrossRef].

Chidananda Kashyap, Bharath Y Kashyap, Guruprasad K, Shrinivasa and D, Prashant Kumar Shrivastava, 2019. Recent Development of Automation and IoT in Agriculture. International Journal of Recent Technology and Engineering (IJRTE) ISSN: 2277-3878, Volume-8, Issue-2S3, July 2019.

Cseri, A, L. Sass, O. Törjék, J. Pauk, I. Vass, and D. Dudits, 2013. "Monitoring drought responses of barley genotypes with semi-robotic phenotyping platform and association analysis between recorded traits and allelic variants of some stress genes," Australian Journal of Crop Science, Vol. 7, no. 10, pp. 1560–1570.

Department for Environment, Food and Rural Affairs, 2016. Water quality and agriculture: Basic Measures. Impact Assessment. 16 Jun 2016 [Online]. Available: http://www.legislation.gov.uk/ukia/2018/27/pdfs/ukia_20180027_en.pdf.

Dursun, M and Ozden, S., 2011. A wireless application of drip irrigation automation supported by soil moisture sensors. Sci. Res. Essays 6 (7), 1573–1582.

Eben-Chaime, M., Bechar, A., and Baron, A. 2011. Economical evaluation of greenhouse layout design. International Journal of Production Economics, 134(1), 246-254.

EcoRobotix, 2019. A Solar-Powered Autonomus Weeding Robot by Sia Magazine - July 21.

Eizicovits, D and Berman, S. 2014. Efficient sensory-grounded grasp pose quality mapping for gripper design and online grasp planning. Robotics and Autonomous Systems, 62(8), 1208-1219.

Elsevier B.V, 2019. Artificial Intelligence in Agriculture 2. This is an open access article under the CC BY-NC-ND license Contents lists available at Science Direct journal homepage: http://www.keaipublishing.com/en/journals/

artificialintelligence-in-agriculture/(http://creativecommons.org/licenses/by-nc-nd/4.0/).

Emmi, L.; Gonzalez-De-Soto, M.; Pajares, G and Gonzalez-De-Santos, P. 2014. Integrating Sensory/Actuation Systems in Agricultural Vehicles. Sensors 14, 4014–4049. [CrossRef]

Emmi, L.; Gonzalez-De-Soto, M.; Pajares, G and Gonzalez-De-Santos, P. 2014. New Trends in Robotics for Agriculture: Integration and Assessment of a Real Fleet of Robots. Sci. World J, 2014, 1–21. [CrossRef]

Encinas, C., Ruiz, E., Cortez, J. and Espinoza, A., 2017. Design and implementation of a distributed IoT system for the monitoring of water quality in aquaculture. Wireless Telecommunications Symposium (WTS). 2017, 1–7.

EU Robotics AISBL. 2014. Robotics 2020 multi-annual roadmap for robotics in Europe, call 1 ict23 – horizon 2020. [Online]. Available: https://www.eu-robotics.net/cms/upload/downloads/ppp-documents/Multi-Annual_Roadmap2020_ICT- 24_Rev_B_full.pdf.

European Parliamentary Research Service, 2016. Precision agriculture and the future of farming in Europe. Scientific Foresight Study. [Online]. Available: http://www.europarl. europa. eu/RegData/etudes/STUD/2016/581892/EPRS_ STU(2016)581892_EN.pdf.

European Space Agency. 2018. Sentinel online. [Online]. Available: https://sentinel.esa.int/web/sentinel/home.

EU SCAR. 2016. Agricultural Knowledge and Innovation Systems towards the Future–A Foresight Paper; Publications of the European Union: Luxembourg.

FAO. 2021. Keeping Plant Pests and Diseases at Bay: Experts Focus on Global Measures. Available online: (accessed on 15 February 2021).

Fentanes, J. P, I. Gould, T. Duckett, S. Pearson, and G. Cielniak, 2018. "3d soil compaction mapping through krigingbased exploration with a mobile robot," arXiv preprint arXiv:1803.08069.

Ftobe, 2007. The Ultimate Guide to Agricultural Robotics, January 1, 2017 https://www.roboticsbusinessreview.com › agriculture.

Gerard Sylvester, 2015. Success Stories on Information and Communication Technologies for Agriculture and Rural Development. Food and Agriculture Organization of the United Nations Regional Office for Asia and the Pacific, Bangkok, Trap Publication 2015/02.

Gonzalez-de-Soto, M., Emmi, L., Benavides, C., Garcia, I and Gonzalez-de-Santos, P. 2016. Reducing air pollution with hybrid-powered robotic tractors for precision agriculture. Biosystems Engineering, 143: 79-94.

Gonzalez-De-Santos, P.; Ribeiro, A.; Fernandez-Quintanilla, C.; Lopez-Granados, F.; Brandstoetter, M.; Tomic, S.; Pedrazzi, S.; Peruzzi, A.; Pajares, G and Kaplanis,

G. 2016, Fleets of robots for environmentally-safe pest control in agriculture. Precis. Agric. 18, 574–614. [CrossRef]

Gonzalez-de-Soto, M., Emmi, L., Garcia, I., and Gonzalez-deSantos, P. 2015. Reducing fuel consumption in weed and pest control using robotic tractors. Computers and Electronics in Agriculture, 114: 96-113.

Graves, A., J. Morris, L. Deeks, R. Rickson, M. Kibblewhite, J. Harris, T. Farewell, and I. Truckle. 2015. "The total costs of soil degradation in England and Wales," Ecological Economics, vol 119, pp. 399–413, Nov. 2015 [Online]. Available: https://doi. org/10.1016/j.ecolecon. 2015.07.026.

Grimstad, L and P. From, 2017. "The Thorvald II agricultural robotic system," Robotics, vol. 6, no. 4, p. 24, Sep 2017. [Online]. Available: https://doi. org/10.3390/robotics6040024.

Haddadin, S and E. Croft, 2016. "Physical human–robot interaction," in Springer Handbook of Robotics. Springer International Publishing, pp. 1835–1874. [Online]. Available: https://doi.org/10.1007/978-3-319-32552-1_69.

Haibo L, Dong S, Zunmin, L and Chuijie, Y. 2015. Study and Experiment on a Wheat Precision Seeding Robot. J. Robot., 1, 1–9.

Hansen, M. F, M. L. Smith, L. N. Smith, M. G. Salter, E. M. Baxter, M. Farish, and B. Grieve, 2018. "Towards on-farm pig face recognition using convolutional neural networks," Computers in Industry, vol. 98, pp. 145–152, Jun 2018. [Online]. Available: https://doi.org/10.1016/j.compind.2018.02.016.

Hansen, M, M. Smith, L. Smith, K. A. Jabbar, and D. Forbes, 2018. "Automated monitoring of dairy cow body condition, mobility and weight using a single 3d video capture device," Computers in Industry, vol. 98, pp. 14–22, Jun 2018. [Online]. Available: https://doi.org/10.1016/j.compind.2018.02.01.

Haug, S, A. Michaels, P. Biber, and J. Ostermann, 2014. "Plant classification system for crop/weed discrimination without segmentation," in IEEE Winter Conference on Applications of Computer Vision. IEEE, Mar 2014. [Online]. Available: https://doi.org/10.1109/wacv. 2014.6835733.

Hayashi, S, Yamamoto, S, Saito, S, Ochiai, Y, Kamata, J, Kurita, M and Yamamoto, K. 2014. Field Operation of a Movable Strawberry-harvesting Robot using a Travel Platform. Jpn. Agric. Res. Q. 48, 307–316.

Hayhow, D.B. 2016. State of Nature. The State of Nature partnership. [Online]. Available: https://www.rspb.org.uk/our-work/stateofnature2016/.

Heba Soffar, 2016. Agricultural robots advantages and disadvantages Published May 11, 2016 · Updated February 25, 2020 https://www.online-sciences.com › robotics › agricultur.

Hiremath, Van der Heijden, G. W. A. M., Van Evert, F. K., Stein, A., and Ter Braak, C. J. F. 2014. Laser range finder model for autonomous navigation of a robot in a maize field using a particle filter. Computers and Electronics in Agriculture, 100, 41-50.

Hitz, G, F. Pomerleau, M.-E. Garneau, C. Pradalier, T. Posch, J. Pernthaler, and R. Siegwart, 2012. "Autonomous inland water monitoring: Design and application of a surface vessel," IEEE Robotics and Automation Magazine, vol. 19, no. 1, pp. 62–72, Mar 2012. [Online]. Available: https://doi.org/10.1109/mra.2011.2181771.

HM Government, 2017. Industrial strategy: Building a Britain fit for the future. [Online]. Available 27 Nov 2017: https://www.gov. uk/government/publications/industrial-strategy-buildingabritain- fit-for-the-future.

Hong Y, JeonLei and F.Tian. 2009. Direct application end effector for a precise weed control robot Author links open overlay panel. Biosystems Engineering, 104(4) 458-464.

Ingale, H.T. and Kasat, N.N., 2012. Automated irrigation system. Int. J. Eng. Res. Dev. 4 (11), 51–54.

International Federation of Robotics, 2017. Executive summary world robotics 2017 service robots. [Online]. Available: https://ifr.org/downloads/press/Executive_Summary_WR_ Service_Robots_2017_1.pdf.

Jabbar, K. A, M. F. Hansen, M. L. Smith, and L. N. Smith, 2017. "Early and non-intrusive lameness detection in dairy cows using 3-dimensional video," Biosystems Engineering, vol. 153, pp. 63–69, Jan 2017. [Online]. Available: https://doi.org/10.1016/j. biosystemseng.2016.09.017.

Jie, L.; Jiao, S.; Wang, X. and Wang, H. 2019. A new type of facility strawberry stereoscopic cultivation mode. J. China Agric. Univ. 24, 61–68.

Jpvalery, 2018. Is the Future of Agriculture in Robotics? Posted on 17/05/2018 https://www.robotshop.com › community › blog › show.

Kalaivani, T, Allirani, A and Priya, P. 2011. A survey on Zigbee based wireless sensor networks in agriculture. IEEE 85–89.

Kamata, T. A. and Roshanianfard, N. N. 2018. "Heavy-weight Crop Harvesting Robot-Controlling Algorithm," IFAC-PapersOnLine, vol. 51, no. Issue 17, p. Pages 244-249, 2018.

Kamelia, L, M. A. Ramdhani, A. Faroqi, and V. Rifadiapriyana, 2018. "Implementation of Automation System for Humidity Monitoring and Irrigation System," IOP Conf. Ser. Mater. Sci. Eng., vol. 288, no. 1, p. 012092, Jan.

Kamilaris, A and F. X. Prenafeta-Boldú, 2018. "Deep learning in agriculture: A survey," Computers and Electronics in Agriculture, vol. 147, pp. 70–90, Apr 2018. [Online]. Available: https://doi. org/10.1016/j.compag.2018.02.016.

Katariya, S.S., Gundal, S.S., Kanawade, M.T and Mazhar, K., 2015. Automation in agriculture. International Journal of Recent Scientific Research. 6 (6), 4453–4456.

Kayacan, E.; Ramon, H and Saeys, W, 2015. Towards agrobots: Identification of the yaw dynamics and trajectory tracking of an autonomous tractor. Comput. Electron. Agric. 115, 78–87. [CrossRef]

Khan, N, Medlock, G, Graves, S and Anwar, S. 2018. GPS Guided Autonomous Navigation of a Small Agricultural Robot with Automated Fertilizing System; SAE Technical Paper; SAE International: Warrendale PA, USA,; Volume 1.

Kim, S, C. Laschi, and B. Trimmer, 2013. "Soft robotics: a bioinspired evolution in robotics," Trends in Biotechnology, vol. 31, no. 5, pp. 287–294, May 2013. [Online]. Available: https://doi.org/10.1016/j.tibtech.2013.03.002.

Kongskilde. 2017. New Automated Agricultural Platform–Kongskilde Vibro Crop Robotti. 2017. Available online: http://conpleks.com/robotech/new-automated (accessed on 2 July2017).

Kozai, T, G. Niu, and M. Takagaki, 2015. Plant Factory: An Indoor Vertical Farming System for Efficient Quality Food Production. Academic Press, 2015.

Kumar, G., 2014. Research paper on water irrigation by using wireless sensor network. International Journal of Scientific Engineering and Technology 123–125 IEERT conference Paper.

Kumba Sennaar, 2019. Agricultural Robots – Present and Future Applications. Last updated on February 3, 2019. https://emerj.com › ai-sector-overviews › agricultural-r.

Kusumam, K, T. Krajník, S. Pearson, T. Duckett, and G. Cielniak, 2017. "3d-vision based detection, localization, and sizing of broccoli heads in the field," Journal of Field Robotics, vol. 34, no. 8, pp. 1505–1518, Jun 2017. [Online]. Available: https://doi. org/10.1002/rob.21726.

Lakhwani, K, H. Gianey, N. Agarwal, and S. Gupta, 2019. "Development of IoT for Smart Agriculture a Review," Springer, Singapore. pp. 425–432.

Lee, W.S and Slaughter, D.C. 1998. Plant recognition using hardware-based neural network. In Proceedings of the 1998 ASAE Annual International Meeting, Orlando, FL, USA, 12–16 July; pp. 1–14.

Lee, W.S, Slaughter, D.C and Giles, D.K. 1999. Robotic Weed Control System for Tomatoes. Precis. Agric. 1, 95–113.

Levin, M. A. D. 2016, "Design of a Task-Based Modular Re-Configurable Agricultural Robot," IFAC-PapersOnLine, vol. 49, no. Issue 16, p. Pages 184-189, 2016.

Logatchevl, A.A., Afanas'evl, V.P., Shkoltnik, S.M., Pursch, H and Jiittner, B., 1998. The behavior of vacuum arc discharges at hydrogen impregnated electrodes. IEEE 18th Int. Symp. on Discharses and Electrical Insulation in Vacuum-Eindhoven, pp. 288–291.

Lottes, P, M. Hörferlin, S. Sander, and C. Stachniss, 2016. "Effective vision-based classification for separating sugar beets and weeds for precision farming,"

Journal of Field Robotics, vol. 34, no. 6, pp. 1160–1178, Sep 2016. [Online]. Available: https://doi.org/10.1002/rob.21675.

Mandow, A. Gomez-de-Gabriel, J.M, Martinez, J.L, Munoz, V.F, Ollero, A and Garcia-Cerezo, A. 1996. The autonomous mobile robot AURORA for greenhouse operation. IEEE Robot. Autom. Mag, 3, 18–28.

Maier, J., 2017. Made smarter review. Department for Business, Energy and Industrial Strategy. [Online]. Available: http://hdl.voced.edu.au/10707/444094.

Mathiassen, S. K, T. Bak, S. Christensen, and P. Kudsk, 2006. "The effect of laser treatment as a weed control method," Biosystems Engineering, vol. 95, no. 4, pp. 497–505, Dec 2006. [Online]. Available: https://doi.org/10.1016/j.biosystemseng.2006.08.010.

Mekonnen M. M and A. Y. Hoekstra, 2016. "Four billion people facing severe water scarcity," Science Advances, vol. 2, no. 2, pp. e1 500 323–e1 500 323, Feb 2016. [Online]. Available: https://doi.org/10.1126/sciadv.1500323.

Mina Mesbahi, 2021. The Future of Agriculture: Adopting More Advanced Robots for Next-Gen Farming. 18 Jan, 2021. https://www.wevolver.com › article › the-future-of-agri.

Mohanraj, I., Ashokumar, K. and Naren, J., 2016. Field monitoring and automation using IoT in agriculture domain. 6th International Conference on Advances in Computing and Communications.

Nagasaka, Y.; Saito, H.; Tamaki, K.; Seki, M.; Kobayashi, K and Taniwaki, K. 2009. An autonomous rice transplanter guided by global positioning system and inertial measurement unit. J. Field Robot. 26, 537–548. [CrossRef]

Naïo Technologies, 2020. Multifunctional Vineyard Weeding Robot–TED. Available online: https://www.naioTechnologies.com/en/agricultural-equipment/vineyard-weeding-robot/(accessed on 2 July 2020).

National Farmers Union report. 2017. Contributions of UK agriculture. [Online]. Available: https://www.nfuonline.com/assets/93419. Feb 2017.

Nawaz, M, Bourrié, G and Trolard, F. 2012. Soil compaction impact and modelling: A review. Agron. Sustain. Dev. 33.

New Holland. 2020. Available online: http://agriculture1.newholland.com/nar/en-us/about-us/whats-up/newsevents/2016/new-holland-nh-drive-new-concept-autonomous-tractor (accessed on 2 July 2020).

Ng, K. C and Trivedi, M. M. 1998. A neuro-fuzzy controller for mobile robot navigation and multirobot convoying. IEEE Transactions on Systems, Man, and Cybernetics, Part B: Cybernetics, 28(6), 829-840.

Nishiura, Y., Y. Yasukuri, and J. Hanamura, 2019. Robotics and System Technologies for Raising and Grafting Seedlings. IFAC Proceedings Volume 46.

Nof, S. Y. 2009. Handbook of automation (1st ed.). Springer.

Noguchi, N. O. B. J. 2011, "Robot farming system using multiple robot tractors in Japan agriculture," IFAC Proc. Vol., vol. Volume 44, no. Issue 1, p. Pages 633-637, 2011.

Noguchi, N, Reid, J, Benson, E and Stombaugh, T. 1998. Vision Intelligence for an Agricultural Mobile Robot Using a Neural Network. IFAC Proc. Vol, 31, 139–144.

Noguchi, N, Reid, J.F, Ishii, K and Terao, H. 2001. Multi-Spectrum Image Sensor for Detecting Crop Status by Robot Tractor. IFAC Proc. Vol, 34, 111–115.

Noguchi, N.; Reid, J.F.;Will, J.; Benson, E.R and Stombaugh, T.S. 1998. Vehicle automation system based on multi-sensor integration. In Proceedings of the 1998 Annual International Meeting (ASAE), Orlando, FL, USA, 12–16 July 1998; Paper No. 983111. pp. 49085–49659.

Noguchi, N., Will, J., Reid, J., and Zhang, Q. 2004. Development of a master-slave robot system for farm operations. Computers and Electronics in Agriculture, 44(1), 1-19.

Nørremark, M.; Griepentrog, H.; Nielsen, J and Søgaard, H. 2008. The development and assessment of the accuracy of an autonomous GPS-based system for intra-row mechanical weed control in row crops. Biosyst. Eng. 101, 396–410. [CrossRef]

Oberti, R., M. Marchi, P. Tirelli, A. Calcante, M. Iriti, M. Hoèevar, J. Baur, J. Pfaff,C. Schütz, and H. Ulbrich. 2013. Journal of Agricultural Engineering.154 (29).

O'Connor, M.; Bell, T.; Elkaim, G and Parkinson, B. 1996. Automatic Steering of Farm Vehicles Using GPS. In Proceedings of the 3rd International Conference on Precision Agriculture, Minneapolis, MN, USA, 23–26 June 1996; pp. 767–777.

Olsen, D. R and S. B. Wood, 2004. "Fan-out: Measuring human control of multiple robots," in Proceedings of the 2004 conference on Human factors in computing systems (CHI). ACM Press. [Online]. Available: https://doi.org/10.1145/985692.985722.

Patel, P.N.; Patel, M.A.; Faldu, R.M and Dave, Y.R. 2013. Quadcopter for Agricultural Surveillance. Adv. Electron. Electr. Eng. 3, 427–432.

Patil, S.S and Thorat, S.A., 2016. Early detection of grapes diseases using machine learning and IoT. Second International Conference on Cognitive Computing and Information Processing (CCIP), IEEE 1–5.

Peleg, D. 2005. Distributed Coordination Algorithms for Mobile Robot Swarms: New Directions and Challenges. In Proceedings of the Distributed Computing–IWDC 2005, Kharagpur, India, 27–30 December 2005 Springer: Berlin/Heidelberg, Germany, 2005; pp. 1–12.

Pereira A and M. Althoff, 2017. "Over-approximative human arm occupancy prediction for collision avoidance," IEEE Transactions on Automation Science and Engineering, vol. 15, no. 2, pp. 818–831, Apr 2018. [Online]. Available: https://doi.org/10.1109/tase.2017.2707129.

Pilarski, T.; Happold, M.; Pangels, H.; Ollis, M.; Fitzpatrick, K and Stentz, A. 2002. The Demeter System for Automated Harvesting. Auton. Robot. 13, 9–20. [CrossRef]

Pilli, S.K, Nallathambi, B, George, S.J and Diwanji, V. 2015. eAGROBOT–A robot for early crop disease detection using image processing. In Proceedings of the 2015 2nd International Conference on Electronics and Communication Systems (ICECS), Coimbatore, India, 26–27 February 2015; pp. 1684–1689.

Piper Thomson, 2019. Types of Robots Shaping Our World. August 26, 2019. https://learn.g2.com › types-of-robots.

Potts, S.P. and Neumann, B.V.S. 2018, "Robotic bees for crop pollination: Why drones cannot replace biodiversity," Sci. Total Environ. vol. 642, p. 665-667.

Pound, M. P, J. A. Atkinson, A. J. Townsend, M. H. Wilson, M. Griffiths, A. S. Jackson, A. Bulat, G. Tzimiropoulos, D. M. Wells, E. H. Murchie, T. P. Pridmore, and A. P. French, 2017. "Deep machine learning provides state-of-the-art performance in image-based plant phenotyping," GigaScience, vol. 6, no. 10, pp. 1–10, Aug 2017. [Online]. Available: https://doi.org/10.1093/gigascience/gix083.

Precision Makers Greenbot, 2020. Available online: http://www.precisionmakers.com/greenbot/(accessed on 2 July 2020).

Qin Zhang, 2013. Agricultural Automation, Fundamentals and Practices 1st Edition Edited By Francis J. Pierce Copyright Year 2013. https://www.automate.org › blogs › robotics-in-agricult.)

Raussendorf 2020. Raussendorf Maschinen. Available online: http://www.raussendorf.de/en/fruit-robot.html (accessed on 2 July 2020).

Robotics in Agriculture, 2017. Types and Applications, Robotics Online Marketing Team.

Ruckelshausen, A.; Biber, P.; Dorna, M.; Gremmes, H.; Klose, R.; Linz, A.; Rahe, R.; Resch, R.; Thiel, M and Trautz, D. 2009. BoniRob: An autonomous field robot platform for individual plant phenotyping. Precis. Agric., 9, 841–847.

Sa, I, Z. Ge, F. Dayoub, B. Upcroft, T. Perez, and C. McCool, 2016. "Deep Fruits: A fruit detection system using deep neural networks," Sensors, vol. 16, no. 8, p. 1222, Aug 2016. [Online]. Available: https://doi.org/10.3390/s16081222.

Sakaue, O. 1996. Development of seeding production robot and automated transplanter system. Jpn. Agric. Res. Q, 30, 221–226.

Sammons, Phil, Furukawa, Tomonari, Bulgin and Andrew. 2005. Autonomous Pesticide Spraying Robot for use in a Greenhouse.

Schor, N, Bechar, A, Ignat, T, Dombrovsky, A, Elad, Y and Berman, S. 2016. Robotic Disease Detection in Greenhouses: Combined Detection of Powdery Mildew and Tomato Spotted Wilt Virus. IEEE Robot. Autom. Lett., 1, 354–360.

Schueller, J. K. 2006. CIGR handbook of agricultural engineering (Vol. VI). CIGR e The International Commission of Agricultural Engineering.

Shahin, M.A. and Symons, S.J., 2001. A machine vision system for grading lentils. Can. Biosyst. Eng. 43, 7–14.

Shekhar, Y., Dagur, E., Mishra, S., Tom, R.J., Veeramanikandan, M and Sankaranarayanan, S. 2017. Intelligent IoT based automated irrigation system. Int. J. Appl. Eng. Res. 12 (18), 7306– 7320.

Shepherd, R. F, F. Ilievski, W. Choi, S. A. Morin, A. A. Stokes, A. D. Mazzeo, X. Chen, M. Wang, and G. M. Whitesides, 2011. "Multigait soft robot," Proceedings of the National Academy of Sciences, vol. 108, no. 51, pp. 20 400–20 403, Nov 2011. [Online]. Available: https://doi.org/10.1073/pnas.1116564108.

Siciliano, B and Khatib, O. 2016. Springer Handbook of Robotics, 2nd ed.; Springer Publishing Company: Cham, Switzerland.

Sinden, J.A. 2004. Australian Weed Management (Australia), C.R.C. The Economic Impact of Weeds in Australia: Report to the CRC for Australian Weed Management; CRC Weed Management: Adelaide, Australia, p. 55.

Sirisha, V and D. Somasekhar, 2019. "Design and Development of Agriculture System Based on IoT."

Sistler, F. 1987. Robotics and intelligent machines in agriculture. IEEE J. Robot. Autom., 3, 3–6. Raussendorf. Fruit Robot. Available online: (accessed on 1 March 2021).

Slaughter, David, Giles, D.K. and Downey, D. 2008. Autonomous robotic weed control systems: A review. Computers and Electronics in Agriculture. 61. 63-78.

Smith, L. N, W. Zhang, M. F. Hansen, I. J. Hales, and M. L. Smith, 2018. "Innovative 3d and 2d machine vision methods for analysis of plants and crops in the field," Computers in Industry, vol. 97, pp. 122–131, May 2018. [Online]. Available: https://doi. org/10.1016/j.compind.2018.02.002.

Sneha, K., R. Kamath, M. Balachandra, and S. Prabhu, 2019. "New Gossiping Protocol for Routing Data in Sensor Networks for Precision Agriculture," Springer, Singapore, pp. 139–152.

Sørensen C and D. Bochtis, 2010. "Conceptual model of fleet management in agriculture," Biosystems Engineering, vol. 105, no. 1, pp. 41–50, Jan 2010. [Online]. Available: https://doi. org/10.1016/j.biosystemseng.2009.09.009.

Stentz, A.; Dima, C.; Wellington, C.; Herman, H and Stager, D. 2002, A System for Semi-Autonomous Tractor Operations. Auton. Robot. 13, 87–104 [CrossRef].

STFC Research Grant ST/N006836/1, 2016. Synthesis of remote sensing and novel ground truth sensors to develop high resolution soil moisture forecasts in China and the UK. [Online]. Available: http://gtr.ukri.org/projects?ref=ST per cent 2FN006836 per cent 2F1.

Subramanian, V.; Burks, T.F and Arroyo, A. 2006, Development of machine vision and laser radar based autonomous vehicle guidance systems for citrus grove navigation. Comput. Electron. Agric. 53, 130–143. [CrossRef].

Sylvester, G. 2018. E-Agriculture in Action: Drones for Agriculture; Food and Agriculture Organization of the United Nations and International Telecommunication Union: (Ed.) Bangkok, Thailand.

Tabile, R and E. Godoy. 2010, "Design of the mechatronic architecture of an agricultural mobile robot," IFAC Proc., vol. Volume 43, no. Issue 18, p. Pages 717-724.

Tamaki, K., Y. Nagasaka and K. Nishiwaki, 2013, "A robot system for paddy field farming in Japan," IFAC Proc. Vol., vol. 46, no. Issue 18, pages 143-147.

Tan, L., 2016. Cloud-based decision support and automation for precision agriculture in orchards. IFAC-Papers OnLine. 49 (16), 330–335.

Tassou, S. A, M. Kolokotroni, B. Gowreesunker, V. Stojceska, A. Azapagic, P. Fryer, and S. Bakalis, 2014. "Energy demand and reduction opportunities in the UK food chain," Proceedings of the Institution of Civil Engineers - Energy, vol. 167, no. 3, pp. 162–170, Aug 2014. [Online]. Available: https://doi. org/10.1680/ener.14.00014.

Thuilot, B.; Cariou, C.; Cordesses, L and Martinet, P. 2001. Automatic guidance of a farm tractor along curved paths, using a unique CP-DGPS. In Proceedings of the 2001 IEEE/RSJ International Conference on Intelligent Robots and Systems. Expanding the Societal Role of Robotics in the Next Millennium (Cat. No.01CH37180), Maui, HI, USA, 29 October–3 November 2001; Institute of Electrical and Electronics Engineers (IEEE): New York, NY, USA, Volume 2, pp. 674–679.

Tijmen Bakker, Kees van Asselt, Jan Bontsema, Eldert J and Van Henten, 2010. Agricontrol, IFAC International Conference, 6-8 December, 2010, Kyoto, Japan.

Tillett, N, T. Hague, A. Grundy, and A. Dedousis, 2008. "Mechanical within-row weed control for transplanted crops using computer vision," Biosystems Engineering, vol. 99, no. 2, pp. 171– 178, Feb 2008. [Online]. Available: https://doi. org/10.1016/j.biosystemseng.2007.09.026.

Tremblay, N., Bouroubi, M.Y., Panneton, B., Guillaume, S., Vigneault, P., and Be'lec, C., 2010. Development and validation of fuzzy logic inference to determine optimum rates of N for corn on the basis of field and crop features. Precision Agric. 11, 621–635.

Tremblay, N., Fallon, E., and Ziadi, N. 2011. Sensing of crop nitrogen status: Opportunities, tools, limitations, and supporting information requirements. Horttechnology, 21(3), 274-281.

Tremblay, N., Wang, Z. J., and Cerovic, Z. G. 2012. Sensing crop nitrogen status with fluorescence indicators. A review. Agronomy for Sustainable Development, 32(2), 451-464.

Tsolakis, N.; Bechtsis, D and Bochtis, D. 2019. AgROS:A Robot Operating System Based Emulation Tool for Agricultural Robotics. Agronomy 9, 403. [CrossRef]

Underwood, J.P.; Calleija, M.; Taylor, Z.; Hung, C.; Nieto, J.; Fitch, R and Sukkarieh, S. 2015 Real-time target detection and steerable spray for vegetable crops. In Proceedings of the IEEE International Conference on Robotics and Automation (ICRA)–Workshop on Robotics in Agriculture, Seattle, WA, USA, 25–30 May 2015; pp. 504–510.

Urrea, C., and Munoz, J. 2015. Path tracking of mobile robot in crops. Journal of Intelligent and Robotic Systems, 80(2): 193-205.

Utstumo, T., Urdal, F., Brevik, A., Dørum, J., Netland, J., Overskeid, Ø., Berge, T.W. and Gravdahl, J.T., 2018. Robotic in-row weed control in vegetables. Computers and Electronics in Agriculture, 154: 36-45.

Walter, A, R. Khanna, P. Lottes, C. Stachniss, R. Siegwart, J. Nieto, and F. Liebisch, 2018. "Flourish - a robotic approach for automation in crop management," in Proceedings of the International Conference on Precision Agriculture (ICPA).

Wolfert, S, L. Ge, C. Verdouw, and M.-J. Bogaardt, 2017. "Big data in smart farming – a review," Agricultural Systems, vol. 153, pp. 69–80, May 2017. [Online]. Available: https://doi. org/10.1016/j.agsy.2017.01.023.

Xia, C., Wang, L., Chung, B.-K and Lee, J.-M. 2015. In situ 3D segmentation of individual plant leaves using a RGB-D camera for agricultural automation. Sensors, 15(8), 20463- 20479.

Xue, J., L and Zhang, T. G. 2012, "Variable field-of-view machine vision based row guidance of an agricultural robot," Comput. Electron. Agric., vol. Volume 84, p. Pages 85-91, 2012.

Yong,W., Shuaishuai, L., Li, L., Minzan, L., Arvanitis, K.G., Georgieva, C., and Sigrimis, N., 2018. Smart sensors from ground to cloud and web intelligence. IFAC-Papers OnLine 51 (17), 31–38./Artificial Intelligence in Agriculture 2 (2019) 1–12.

Zhang, C. and J. M. Kovacs, 2012. "The application of small unmanned aerial systems for precision agriculture: a review," Precision Agriculture, vol. 13, no. 6, pp. 693–712, jul 2012. [Online]. Available: https://doi.org/10.1007/s11119-012-9274-5.

Zhang J. 2016. "Field phenotyping robot design and validation for the crop breeding," IFAC-PapersOnLine, vol. Volume 49, no. Issue 16, p. Pages 281-286,

Zhang, Q. 2013. Opportunity of robotics in specialty crop production. IFAC Proceedings Volumes, 46(4), 38-39.

Zhang, Z., Noguchi, N., Ishii, K., Yang, L., and Zhang, C. 2013. Development of a robot combine harvester for wheat and paddy harvesting. IFAC Proceedings Volumes, 46(4), 45-48.

Zion, B., Mann, M., Levin, D., Shilo, A., Rubinstein, D., and Shmulevich, I. 2014. Harvest order planning for a multiarm robotic harvester. Computers and Electronics in Agriculture, 103, 75-81. Ludhiana, Punjab. See discussions, status and author profiles for this publication at: https://www.researchgate.net/publication/342096836.

Index

9 789359 190921

Printed at Repro India Ltd.